The European Union and Global Environmental Protection

This book examines how the EU can be a more proactive actor in the promotion of the principles of sustainability and fairness from a legal environmental perspective. The book is one of the results of the research activity of the Jean Monnet Chair in EU Environmental Law (2017–2020), funded by the European Commission under the Erasmus+ programme.

The European Union and Global Environmental Protection: Transforming Influence into Action begins with an introduction of the key EU competences, instruments and mechanisms, as well as the current international challenges at the EU level. It then explores case study examples from four regulated fields: climate change, biodiversity, multilateral trade, unregulated fishing, and access to justice; and four unregulated areas: migratory issues, environmental crime, mainstreaming of the Sustainable Development Goals in EU policies, and environmental justice, highlighting the extent to which the EU might align with international environmental regimes or extend its normative power.

This volume will be of great relevance to students, scholars, and EU policy makers with an interest in international environmental law and policy.

Mar Campins Eritja is Full Professor of Public International Law at the University of Barcelona. Her main areas of research are European and international environmental law. She leads and participates in various competitive research projects and has published extensively in these areas. She is currently the holder of the UB Jean Monnet Chair on EU Environmental Law.

Routledge Studies in Environmental Policy

Strategic Designs for Climate Policy Instrumentation
Governance at the Crossroads
Gjalt Huppes

The Right to Nature
Social Movements, Environmental Justice and Neoliberal Natures
Edited by Elia Apostolopoulou and Jose A. Cortes-Vazquez

Guanxi and Local Green Development in China
The Role of Entrepreneurs and Local Leaders, 1st Edition
Chunhong Sheng

Environmental Policy in India
Edited by Natalia Ciecierska-Holmes, Kirsten Jörgensen, Lana Ollier and D. Raghunandan

Mainstreaming Solar Energy in Small, Tropical Islands
Cultural and Policy Implications
Kiron C. Neale

EU Environmental Governance
Current and Future Challenges
Edited by Amandine Orsini and Elena Kavvatha

The European Union and Global Environmental Protection
Transforming Influence into Action
Edited by Mar Campins Eritja

Environmental Policy and Air Pollution in China
Governance and Strategy
Yuan Xu

For more information about this series, please visit: www.routledge.com/Routledge-Studies-in-Environmental-Policy/book-series/RSEP

The European Union and Global Environmental Protection

Transforming Influence into Action

Edited by Mar Campins Eritja

Routledge
Taylor & Francis Group

LONDON AND NEW YORK

earthscan
from Routledge

First published 2021
by Routledge
2 Park Square, Milton Park, Abingdon, Oxon OX14 4RN

and by Routledge
52 Vanderbilt Avenue, New York, NY 10017

Routledge is an imprint of the Taylor & Francis Group, an informa business

British Library Cataloguing-in-Publication Data
A catalogue record for this book is available from the British Library

Library of Congress Cataloging-in-Publication Data
A catalog record for this book has been requested

ISBN: 978-0-367-89321-7 (hbk)
ISBN: 978-1-003-01851-3 (ebk)

Typeset in Times New Roman
by Apex CoVantage, LLC

This book is published with the support of the Erasmus+ Programme of the European Union under the Jean Monnet Action: Jean Monnet Chair in European Union Environmental Law (2017–2020, Project number – 587220-EPP-1-2017-1-ES-EPPJMO-CHAIR), held by Prof. Mar Campins Eritja. The European Commission's support for the production of this publication does not constitute an endorsement of the contents, which reflect the views only of the authors, and the European Commission cannot be held responsible for any use which may be made of the information contained therein.

Co-funded by the
Erasmus+ Programme
of the European Union

Contents

List of tables ix
List of contributors x
Foreword xiii
List of acronyms xix

1 Introduction 1
MAR CAMPINS ERITJA

2 Some challenges for strengthening the EU's international
influence in the climate change regime 12
MAR CAMPINS ERITJA

1. Introductory remarks 12
2. The shared exercise of EU competences 14
3. A comprehensive and integrated approach to fight climate
change 16
4. The United States' withdrawal from the Paris Agreement and
the EU's stronger alliance with China 19
5. The Brexit and its impact on EU climate policy 21
6. Final remarks 24

3 The EU diplomacy for biodiversity 35
TERESA FAJARDO DEL CASTILLO

1. Introductory remarks 35
2. A EU competence and a legal mandate to protect biodiversity
globally 37
3. Is there a specific EU diplomacy for biodiversity? 39
4. Recommendations for the EU 2030 Biodiversity
Strategy: aligning its objectives with those of Green
Multilateralism 43
5. Final remarks 45

4   **The role of the EU in the promotion of sustainable development through multilateral trade**   50

XAVIER FERNÁNDEZ-PONS

1. *Introductory remarks 50*
2. *The EU's Generalised Scheme of Preferences (GSP) and its special incentive arrangement for sustainable development and good governance (GSP+) 52*
3. *The inclusion of a chapter on sustainable development in the new regional trade agreements (RTAs) concluded by the EU with third countries 53*
4. *The EU and sustainable public procurement 55*
5. *The recent reform of the EU regulation of anti-dumping measures and the issue of so-called social and ecological dumping 56*
6. *EU's trade measures related to carbon footprint 60*
7. *Final remarks 63*

5   **The EU's global leadership in the fight against illegal, unreported, and unregulated fishing**   73

XAVIER PONS RAFOLS

1. *Introductory remarks 73*
2. *The global phenomenon of IUU fishing 73*
3. *International legal concept of IUU fishing 75*
4. *Basis and scope of EU regulations on IUU fishing 76*
5. *Core content of the EU Regulations on IUU fishing 78*
6. *The international effect of EU action: establishment of a list of non-cooperating third countries 79*
7. *The multilateral influence of EU action: the port state measures agreement and other international legal and institutional frameworks to combat IUU fishing 81*
8. *Final remarks 83*

6   **The international dimension of the EU on access to justice in environmental matters**   90

ALEXANDRE PEÑALVER-CABRÉ

1. *Introductory remarks 90*
2. *The role of the EU in strengthening access to justice in environmental matters under international law 90*
3. *The EU's inadequate implementation of the Aarhus Convention on access to justice in environmental matters 93*

4. *The Aarhus Convention Compliance Committee case ACCC/C/2008/32 and the shortcomings in access to justice in environmental matters at the EU level 98*
5. *EU response to the meeting of the parties of Aarhus Convention in case ACCC/C/2008/32 99*
6. *Final remarks 101*

7 **Environmental refugees: reshaping the borders of migration in the EU**   109

SUSANA BORRÀS-PENTINAT

1. *Introductory remarks 109*
2. *People moving in the context of environmental change 110*
3. *Environmental refugees: the forgotten migrants in the EU migration policy 113*
4. *Towards the recognition of environmental refugees at the EU level 117*
    The initial attempt of the European Parliament 117
    The cautious approach of the European Commission and the EU high representative 119
    The voluntary (im)prudence of the European Council 122
5. *Mind the gap: defending and protecting environmental refugees at the EU level 123*
6. *Final remarks 126*

8 **Environmental crime: assessing and enhancing EU compliance with international environmental law**   132

MARIA MARQUES-BANQUE

1. *Introductory remarks 132*
2. *Multilateral environmental agreements, EU law, and criminal sanctions 133*
3. *The Directive 2008/99/EC on the protection of the environment through criminal law 136*
4. *The protection of the environment through criminal law in the EU beyond Directives 2008/99/EC and 2009/123/ EC 138*
5. *A proposal to enable the EU to reinforce its position within the international environmental regime 140*
6. *Final remarks 143*

9   **Mainstreaming Sustainable Development Goals into EU policies**   149
    MAR AGUILERA VAQUÉS

    1. *Introductory remarks 149*
    2. *From the concept of sustainable development to Sustainable
        Development Goals 149*
    3. *Sustainable Development Goals as an EU policy effort 151*
    4. *The EU as a global trailblazer in sustainable
        development 155*
    5. *From mainstreaming Sustainable Development Goals to the
        2019 European Green Deal 156*
    6. *Final remarks 158*

10 **Environmental justice in EU law and policies:
    a fundamental challenge**   166
    JORDI JARIA-MANZANO

    1. *Introductory remarks 166*
    2. *Environmental justice vs sustainable development:
        conceptual considerations about the interpretation of the
        global environmental crisis 167*
    3. *Environmental justice and fundamental EU values 170*
    4. *EU law, EU policies, and environmental justice:
        vulnerability, conflicts, and change 172*
    5. *Final remarks 174*

Index   183

Tables

8.1   MEAs, sanctions, and Directive 2008/99/EC   137

Contributors

Mar Aguilera Vaqués is Associate Professor of Constitutional Law at the University of Barcelona and a founder and co-director of the pro-bono Environmental Law Clinic. She obtained her LL.M. of law from Columbia University in New York. She has been a visiting scholar at Harvard Law School, the European University Institute in Florence, and Université Lumière Lyon 2. She has also served as an international observer and legal expert for the European Union and on United Nations election observation missions. Her main areas of research are constitutional law, rights and freedoms, sustainable development, and environmental protection.

Susana Borràs-Pentinat is Associate Professor of Public International Law and EU Law at Rovira i Virgili University. She obtained her LL.M. in environmental law in 2004. Her main fields of research and publication are climate justice, climate migration, rights of nature, and environmental defenders. She is the director of the master's degree in Environmental Law at Rovira i Virgili University and a member of the Research Group on Environmental Law, Immigration, and Local Government of the Tarragona Centre for Environmental Law Studies. She has been a visiting researcher at a host of international institutions, including the Committee on the Challenges of Modern Society (NATO/CCMS), the Max Planck Institute for Comparative Public Law and International Law, the Center for International Environmental Law, and the Centre for Studies and Research at The Hague Academy of International Law and International Relations. As a legal expert, she has been involved in several international networks, most notably the South American Network for Environmental Migrations, the Earth Law Center – which produced the Earth Law Framework for Marine Protected Areas – and the United Nations Harmony with Nature initiative, and she has served on the advisory board of the Brazilian Institute of Human Rights.

Mar Campins Eritja is Full Professor of Public International Law at the University of Barcelona. Her main areas of teaching and research are EU law and European and international environmental law. She leads and participates in various competitive research projects and has published extensively in these areas. She is currently the holder of the Jean Monnet Chair on EU

Environmental Law at the University of Barcelona and the co-lead researcher of a project funded by the Spanish Ministry of Science. She has been a Fulbright Scholar at the American University Washington College of Law and a visiting professor at the Universities of Ottawa and Montreal, Dalhousie University, Nova Southeastern University, the University of Puerto Rico, and the University of Lapland, in Rovaniemi, Finland. She serves on the board of directors of the UB Water Research Institute, and is a member of the Tarragona Centre for Environmental Law Studies at Rovira i Virgili University and the Spanish Observatory of Environmental Policies. She has been a member of the editorial board of *Revista Catalana de Dret Ambiental* since 2009 and *Revista Catalana de Dret Públic* (2012–2016). Her roles in the University of Barcelona include vice dean at the Faculty of Law (2002–2005 and since 2016) and vice rector and deputy vice rector for international policy (2005–2008), and she currently sits on the Advisory Board of the University of Barcelona.

Teresa Fajardo del Castillo is Associate Professor at the University of Granada, where she teaches public international law in the faculty of law and international and European environmental law in the faculty of sciences. She studied law at the Universities of Granada and Poitiers, and obtained the licence spéciale en droit Européen from the Institut d'Etudes Européennes at the Free University of Brussels. Her research fields are international and European environmental law, EU migration law and policy, and soft law. She is engaged in several inter-university and European Union projects and initiatives relating to international and European environmental law. She has completed a stay at the Legal Service of the European Commission and has been a visiting researcher at the Universities of Brussels, Geneva and Cambridge, and at King's College London.

Xavier Fernández-Pons is Associate Professor of Public International Law at the University of Barcelona. He obtained his PhD in law at the University of Bologna and is a recipient of the diploma from the Centre for Studies and Research of The Hague Academy of International Law. He is a researcher at the Tarragona Centre for Environmental Law Studies at Rovira i Virgili University and a member of the Jean Monnet Chair on EU Environmental Law at the University of Barcelona. His main lines of research are international economic law, the World Trade Organization, regional trade agreements, foreign investment, the European Union's trade policy, trade and the environment, and trade and health, all areas in which he has published extensively.

Jordi Jaria-Manzano is Serra Húnter Fellow of Constitutional and Environmental Law at Rovira i Virgili University. His current research interests include global environmental constitutionalism, Anthropocene law, and environmental justice. He has recently authored *La constitución del Antropoceno* (2020) and jointly edited the *Research Handbook on Global Climate Constitutionalism* (2019) with Susana Borràs.

Maria Marques-Banque is Associate Professor of Criminal Law at Rovira i Virgili University and a researcher at the Tarragona Centre for Environmental Law Studies. Her research interests are focused on environmental crime, transnational crime, EU criminal law, sentencing, and Spanish domestic criminal law. She has been a visiting researcher at the Max Planck Institute for Comparative Public Law and International Law; the Faculty of Law at the University of Technology Sydney, in Australia; the Institute of Marine and Environmental Law at the University of Cape Town; and the Asian-Pacific Centre for Environmental Law at the National University of Singapore. She is currently the director of the Department of Public Law and the director of the Environmental Law Clinic at Rovira i Virgili University.

Alexandre Peñalver-Cabré is Associate Professor of Administrative Law at the University of Barcelona. His main areas of teaching and research are environmental law, public interest law litigation, local government, and EU law. He has led and participated in various research projects, and has published several books and articles. He is the coordinator of the Environmental Law Clinic at the UB and a member of the Jean Monnet Chair on EU Environmental Law at the University of Barcelona and the Observatory of Environmental Policies. He has served on the executive board of the NGO Environmental Defense Fund and is a member of the Environmental Law Alliance Worldwide, as well as a number of research centres (Public Law Institute, Research Group on Local Government, and the Transjus Legal Research Institute). He serves on the editorial board on Spanish case law for *Revista de Administración Pública*. He has been an associate researcher at Harvard University and the director of the Environmental Department of the Olof Palme International Foundation. He has also been a lawyer in environmental public interest litigation for 13 years and a judicial officer for 12 years.

Xavier Pons Rafols is Full Professor of Public International Law at the University of Barcelona. He has participated in several research projects and is currently the co-lead researcher of a project funded by the Spanish Ministry of Science. He is a member of the Jean Monnet Chair on EU Environmental Law at the University of Barcelona. He has also served as an international legal consultant for the FAO. His main areas of research are international law, EU law, international organizations, forestry law, and fisheries law, and he has published articles and books in these areas. At the University of Barcelona, he has served in the Law, Economics and Social Sciences Division as vice president (2001–2002), president (2002–2003), and secretary general (2005–2008), and he is currently the dean of the Faculty of Law (since 2016).

Foreword

This book opens new perspectives. Indeed, while most publications on the external powers of the European Union (EU) in environmental law matters deal in great detail with the question of how the capability for action is divided between the EU and its member states, *Mar Campins Eritja* and her Spanish colleagues take a different approach. They look at specific aspects of international environmental law and policy, and examine if – and to what extent – the EU succeeded in influencing or even steering the evolution of environmental law at an international level. And they treat the EU as a unit, or an autonomous actor in the global discussion, without dwelling on the residual responsibilities of its member states.

This innovative approach, which is rarely found elsewhere in the legal literature, allows the authors to discuss the evolution of international environmental law and policy in selected areas in detail, as well as the practice of the different bilateral and multilateral organizations, bodies, and ad-hoc committees, and point to the difficulties for the EU to advance the elaboration and application of the law. The remarkable feature is that at no place is there made reference to the interests, policies, and laws of the EU member states. Rather, the EU is treated as a completely homogeneous entity, which succeeds or fails to succeed to bring about changes in international environmental law. Such an approach – which was hardly found in the literature on international law until now – was long overdue: as the Lisbon Treaties of 2009 further strengthened the external competence of the EU, it was time to get away from attempts to identify French, German, Spanish, or other interests in the EU's external policy and their influence on the final outcome of the deliberations.

It can only be hoped that the authors' approach in treating the EU as a legal entity of its own in the international negotiations and discussions, and in concentrating on actions rather than on legalistic considerations, will be followed by other authors. Environmental law is well placed to overcome the nation-state discussions in Europe, as the environment does not know frontiers. What it needs is protective action, not legal discussions on competences.

The book's title is promising: it intends to discuss the actions that the EU has undertaken. Actions speak louder than words, we learn from the proverb, and it is necessary and useful to discuss international developments of environmental law with regard to the *actions* that are undertaken.

The different contributions all conclude that the EU does not need to be ashamed of its contribution to the evolution of environmental law. The EU has been active throughout the range of problems to promote political, legal, and human results. Comparisons with nation states or with other regional bodies such as NAFTA, ASEAN, Mercosur, etc. demonstrate how the European continent – the EU being regularly supported in its approach by Norway, Switzerland, or Iceland, and hopefully in future also by the United Kingdom – is able to influence or steer the international legal discussions. No other continent can claim similar efforts – and similar success.

The different authors make it clear, though, that the European influence is different from one law sector to the other. As Campins Eritja points out in her contribution on climate change, the EU is leading the international discussions on climate change by having set precise greenhouse gas emission reduction targets since 1990 and having reached the intended 20 percent reduction, while global emissions increased by some 50 percent since that time, and neither the United States nor China, India, Canada, or Latin American countries succeeded in reducing their emissions at a comparable level. It is thus normal that the instruments used by the EU – the emission allowance trading scheme, the agreements to share the burden among the EU member states, the setting of ambitious reduction targets for 2030 and 2050, the re-orientation of the economy towards renewable sources of energy, etc. – can serve other countries as a model to address climate change issues.

Another example of the EU's leading role in international legal negotiations is given by the very concise presentation of Fernández-Pons on multilateral trade agreements. Fernández-Pons explains, in detail, how the EU endeavours to influence multilateral trade rules by trying to insert environmental requirements in the bilateral and multilateral trade agreements, whenever it is a party of such trade agreements. It leads by giving the example and demonstrates that it is possible to insert and apply conditions on sustainable development in trade agreements. The comparison with other trading nations or bodies makes evident how much the EU is committed to promote the environmental protection of this planet. Neither China nor the United States, neither Mercosur, Japan, nor the Asian countries deploy comparable initiatives to promote sustainable development in trade arrangements.

Pons Rafols describes in detail the largely unknown actions of the EU to reduce illegal, unreported, and unregulated (IUU) fishing, within and outside EU waters. The EU adopted, amongst others, a number of measures with extraterritorial effect, regularly published lists of vessels that are involved in IUU fishing and of countries which do not cooperate to fight such practices, thereby clearly assuming a global leadership in the fight against IUU fishing.

Fajardo del Castillo turns her regard on the EU's green diplomacy and the green diplomatic network to promote the conservation of biodiversity, which is likewise little known. She regrets, though, that a number of very important global diplomatic activities in 2020, supported or initiated by the EU, will have to be postponed due to the corona pandemic. Aguilera Vaqués enchains on that when she retraces the EU efforts to comply with the UN Sustainable Development Goals (SDG) by 2030. She links internal and external measures, with the basic consideration that the EU cannot be a global player when it does not have its own internal system in order. The integration requirement of Article 11 TFEU constitutes a useful legal requirement to re-orient all of the EU's policies towards the achievement of the SDGs, and the Commission's Green Deal 2019 constitutes a first promising step in this regard.

Marques-Banque addresses the difficult area of environmental crime, where few international agreements exist. She concentrates on the duty for the EU to lead by example, despite limited competences in criminal matters, and carefully examines in detail the pros and cons of strengthening the existing provisions on criminal sanctions, underlining that the effective enforcement of international environmental provisions remains an important challenge. I will come back on this issue in a moment.

Of course, the EU does not have solutions to every environmental problem, as the particular contributions by Borràs-Pentinat on ecological migration and by Jaria-Manzano on environmental justice demonstrate. Ecological migration, which occurs mainly in Europe and in America, remains an unresolved problem, as it is caused by environmental conditions in Africa and South and Central America and is strongly influenced by economic or human factors (civil war, terrorism, living conditions, economic policies) and reinforced by the environmental injustice which Jaria-Manzano analyses. As long as the living conditions in industrialized countries and countries in development are as different as at present, there will be ecological migration. The EU's policy towards developing countries, and in particular towards Africa, is not yet, as Borràs-Pentinat rightly explains, oriented to fight ecological migration at source – which would require a re-orientation of the EU's development policy.

Jaria-Manzano discusses the important topic of environmental justice. This concept developed in the United States, where it denounced the fact that dangerous industrial installations and incinerators, landfills or other infrastructure measures, were preferably placed close to poorer quarters of people, thus exposing the poor to greater environmental risk and impairment than the well-off population ("class and race discrimination"). Jaria-Manzano transfers this concept to the global level when he draws the attention to the fact that the "massive exploitation of natural resources" mainly takes place in poorer countries and is based on an "institutional and legal framework essentially biased to produce inequality and injustice".

It is obvious that the standard of living in countries of the world is uneven due to century-long colonization, exploitation of the third world, but also due to climate, soil fertility, corruption, abuse of power, and many other human and natural causes – and also the existence of natural resources. As long as countries do not develop policies to systematically fight at least the human causes of poverty, inequality, and injustice, the support from outside countries or organizations will be of limited – very limited – effect.

Jaria-Manzano sees a way forward for the EU to strengthen the human rights of EU citizens as well as of citizens in third countries. It can only be hoped that he is right and that such a strengthening will lead to a massive increase of support for underprivileged countries by developed countries, with financial support, know-how transfer, trade facilities, education programmes, capacity building, and other means. The coronavirus pandemic has shown that this planet is one and that planetary solutions to health and environmental challenges are necessary. Neither the EU nor other countries – USA, Canada, Australia, China, Japan – are ready at present to accept this inconvenient truth. Thus, until there is a clear shift of policy to reduce environmental and other injustice, ecological migration and environmental injustice will continue to mark the planet's coming years and decades.

Could the EU deploy more efforts to improve global environmental justice? My answer is a clear "yes". For example, the EU has a privileged partnership agreement, the so-called Cotonou-Agreement, with 78 developing states of Africa, the Caribbean, and the Pacific. This Agreement, which already covers climate change issues at present, is to be renewed in 2020; it could be used to develop a "green deal", with pilot and model projects for cooperation and concertation with the partner countries on climate change, biodiversity protection, environmental pollution matters, and other aspects concerning the objectives of the United Nations Sustainable Development Goals. The EU could show that the relationship with third-world countries is based on concepts to overcome environmental, social and economic injustice, and, in particular, poverty, by such an example. It is clear, though, that such a cooperative green deal cannot be successful unless the 78 states and the EU as such have the political will to improve (environmental) justice.

There is one other legal aspect which I raise here, as the different contributions in this book only deal with it accessorily. It is the question of implementation of international environmental agreements. Implementation is the biggest challenge for environmental law, as this section of law has no social group behind it to defend and promote its interests; this is the decisive difference to social law, where trade unions and employers organizations are active in the development of social law; to transport law, where transporters ensure that their interests are not forgotten; to competition law, where competitors very actively influence the elaboration and application of competition law; to energy law, where the big public or private energy undertakings exercise a determining influence on the development of that law; etc., etc.

Environmental law has no such social group behind it. Who would defend the climate, the biodiversity, the protection of oceans or the clean air, or who would try to ensure that international environmental agreements are effectively applied? There are some environmental organizations who stand up for such general interests, but they are dispersed, short of human and financial resources, hampered by linguistic barriers, and exposed to scattered national environmental protection rules.

This structural inequality of environmental law makes the monitoring of its application a first necessity for public authorities, as professional operators have a very limited interest in seeing environmental law effectively applied. The EU developed an internal system of monitoring the application of environmental law, with the Commission as surveillance body, which developed a complex system of reporting obligations, name and shame sanctions, enforcement by the EU Court of Justice, and even, at the end of the day, with financial sanctions against repeated cases of non-compliance.

Peñalver-Cabré describes convincingly how the EU, on the one hand, accepted an implementation and enforcement system under the UNECE Aarhus Convention, which goes much further than almost all other international environmental agreements, but struggles considerably to accept the findings under this mechanism. The decisive point is, though, that the EU started to reform its law in order to comply with the Aarhus Convention by 2021.

International (environmental) law has considerable problems seeing the provisions of the different agreements – which were so solemnly agreed upon – effectively applied, whether it is the Paris Agreement on climate change, the Convention on Biodiversity, or any other agreement. State sovereignty prevents the taking of measures against non-complying states. But the Aarhus Convention, negotiated and agreed with the active support of the EU, shows ways of overcoming this traditional role of state sovereignty. Under that Convention:

- Individuals may cause the investigation of non-compliance by an expert compliance committee
- All information submitted or collected as well as the investigation is public and publicly documented
- Environmental organizations may suggest members of the compliance committee

The EU and all its member states accepted and complied with this system without the slightest reservation, despite the momentary discussions described in the book. Why, then, can such a system not be transferred to other international environmental agreements? Also, the World Trade Organization works with (trade) experts in a similar way. Perhaps tomorrow, the EU might take initiatives to have a similar pattern of compliance examination installed in the Basel Convention on waste transfer, the Biodiversity Convention, or the Convention on Long-range Transboundary Air Pollution. When the 27 member states of the EU and the EU itself accommodate to the compliance procedure under the Aarhus Convention, there is no reason, as Peñalver-Cabré put it, not to expand such a compliance system to other environmental agreements. Such a process might give a voice to the environment, allow countries of the Southern Hemisphere to also raise their concerns, and ensure the long-needed transparency in the application of environmental agreements.

It is high time that new paths are explored in international environmental policy and law, and that new answers are researched about the increasing problems of this planet. The book by Mar Campins Eritja and her colleagues opens such new paths in this search. I hope that its successful reception by the international community will encourage the authors to continue their research in order to further approach the Sustainable Development Goals of the United Nations.

Ludwig Krämer
Derecho y Medio Ambiente, Madrid
2 May 2020

Acronyms

AD	Anti-Dumping Agreement
BIOFIN	UN Biodiversity Financing Initiative
CBD	Convention on Biological Diversity
CBD CoP	Conference of the Parties of the Framework Convention on Biological Diversity
CEAS	Common European Asylum System
CDS	Catch Documentation Schemes
CITES	Convention on the Legal Trafficking of Endangered Species of Wild Llora and Fauna
CFP	Common Fisheries Policy
CLIM	Temporary Committee on Climate Change
CMM	Commercial and Conservation and Management Measures
CoP	Conference of the Parties
CSWD	European Commission's Staff Working Document
EACH-FOR	Environmental Change and Forced Migration Scenarios
EAP	European Environmental Action Plan
EBA	Everything But Arms
EC	European Communities
ECHR	European Convention on Human Rights
EEAS	European External Action Service
EFA	Greens/European Free Alliance Group
EFCA	European Fisheries Control Agency
EGA	Environmental Goods Agreement
EGD	European Green Deal
EU	European Union
EUCJ	EU Court of Justice
EU ETS	EU Emissions Trading Scheme
EUTR	Regulation (EU) No. 995/2010 of the European Parliament and of the Council of 20 October 2010, laying down the obligations of operators who place timber and timber products on the market
FAO	United Nations Food and Agriculture Organization

FLEGT	European Union Program for Forest Law Enforcement, Governance and Trade
GATT	General Agreement on Tariffs and Trade
GDN	Green Diplomacy Network
GDP	Gross Domestic Product
GHG	Greenhouse Gas
GPA	Government Procurement Agreement
GSP	Generalised Scheme of Preferences
GSP+	Generalised Scheme of Preferences + (arrangement for sustainable development and good governance)
HAC	High Ambition Coalition
ICAO	International Civil Aviation Organization
IEEP	Institute for European Environmental Policy
ILO	International Labour Organization
ILUC	Indirect Land Use Change
IMO	International Maritime Organization
IOM	International Organization for Migration
IPBES	Science-Policy Platform on Biodiversity and Ecosystem Services Global Assessment
IPOA-IUU	Plan of Action to Prevent, Deter and Eliminate Illegal, Unreported and Unregulated Fishing
ITLOS	International Tribunal for the Law of the Sea
IUU (fishing)	Illegal, Unreported and Unregulated Fishing
LDC	Least-Developed Countries
MARPOL	International Convention for the Prevention of Pollution from Ships
MDG	Millennium Development Goals
MEA	Multilateral Environmental Agreement
MEP	Members of the European Parliament
MFF	Multiannual Financial Framework
MOFCOM	Ministry of Commerce of China
NDC	Nationally Determined Contributions
NGO	Non-Governmental Organization
NME (countries)	Non-Market Economy (countries)
npr-PPM	Non-Product-Related Processes and Production Methods
OAU	Organization of African Union
OJ	Official Journal
PPMs	Processes and Production Methods
PSMA	Agreement on Port State Measures
RED	Renewable Energy Directive
RFB	Regional Fisheries Bodies
RFMO	Regional Fisheries Management Organizations
RFMA	Regional Fisheries Management Arrangements
RPP	Regional Protection Programmes
RTAs	Regional Trade Agreements

SDG	Sustainable Development Goals
SDSN	Sustainable Development Solutions Network
TEU	Treaty on European Union
TFEU	Treaty on the Functioning of the European Union
TSD	Trade and Sustainable Development
TTIP	Transatlantic Trade and Investment Partnership
UK	United Kingdom
UN	United Nations Organization
UNCCD	United Nations Convention to Combat Desertification
UNCED	UN Conference on Environment and Development
UNCLOS	United Nations Convention on the Law of the Sea
UNEP	United Nations Environmental Programme
UNFCCC	United Nations Framework Climate Change Convention
UNHCR	Office of the United Nations High Commissioner for Refugees
UNTOC	United Nations Convention against Transnational Organized Crime
US	United States of America
UVI	Unique Vessel Identifier
VCLT	Vienna Convention on the Law of Treaties
WTO	World Trade Organization
7EAP	Seventh Environment Programme

1   Introduction

Mar Campins Eritja

This collective book employs an academic approach to analyse the international legal mechanisms through which the European Union (EU) seeks to exert a more proactive impact on the promotion of the principles of sustainability and environmental protection at a global level. To this end, it explores a number of international regimes, some of which are well established while others are still in the process of formation. Within these regimes, the EU aims to play a prominent role in the design of an international environmental governance system, while at the same time addressing issues that are currently the subject of major debate in international forums.

While EU environmental law has had high points and low points, it has clearly been expanding since the mid-1970s. Indeed, EU leadership has been particularly noteworthy in managing past and present environmental challenges. Action taken in recent decades has not only laid the foundations for the current system of EU environmental law, but has also consolidated the EU's role in the framework of international environmental regimes from both the institutional and regulatory perspectives. To this end, the EU has taken action in accordance with the competences attributed to it under the founding Treaties, either expressly or implicitly, and has therefore intervened only in those areas in which the member states have agreed to transfer the exercise of these competences to European institutions. The measures taken thus far by the EU in relation to the protection of the environment have mostly been based on current article 192 of the Treaty on the Functioning of the European Union (TFEU). That said, the adoption and implementation of EU provisions in this area concern matters that fall within the competence of both the EU and the member states, so it is clearly a shared competence whose exercise is subject to two prerequisites, namely the application of the principle of proportionality in relation to the intensity of EU action and the principle of subsidiarity with regard to the assessment of the need for EU intervention. Indeed, article 191 TFEU states that the EU shall only "contribute" to preserving and improving the quality of the environment. Furthermore, article 193 TFEU expressly reserves the right of member states to maintain or adopt measures providing for greater environmental protection.

The normative process of creating EU environmental law has, in fact, responded to the fluidity and evolution of this sector in the international context, to which the EU has made an important contribution by facilitating on many occasions the political and legal commitment needed to find solutions to the challenges of managing the international environmental governance system. Through its participation in international environmental governance mechanisms, the EU has sought coherence between its internal regulatory action and the normative development it promotes at the international level over the years, aiming to serve as a model to be emulated by the international community as a whole. Through these efforts, the EU seeks to become a global benchmark and to reassert its influence as a provider of political objectives and legal techniques beyond its borders.

In this respect, the Seventh EU Environment Action Programme, adopted in November 2013, guides EU action until 2020 and underlines the need "to increase the Union's effectiveness in addressing international environmental and climate-related challenges".[1] This aim covers implementation of the commitments of the United Nations Agenda 2030 and the Sustainable Development Goals (SDGs) adopted in 2015, which are to be achieved not only through the EU's active participation in the shaping of international regimes and processes and the intensification of EU cooperation with third states, but also through better implementation of international environmental law in the territory of its member states. More specifically, one of the thematic priorities of the Seventh Action Programme calls for the EU to become more involved "in existing and new multilateral environmental and other relevant processes, in a more consistent, proactive and effective way".[2] In any case, while the issue is emphasized by the European Commission, the European Parliament has noted the lack of a positive trend or clear progress in the area, and has therefore called for significant improvement with a view to 2050, which will require a strengthened commitment from the member states.

In order to undertake this analysis, the book is structured around nine chapters that examine different sectors or international regimes that have an impact on the environment, posing two key issues that are related to one another: on one hand, the scope of the exercise of the EU's own competences and the EU legal framework and, on the other hand, the EU's more or less active participation in the relevant international tools and mechanisms. Each chapter aims to highlight the extent to which the EU is aligned with the degree of environmental protection envisaged within these sectors and international regimes, or is able to use its normative power to develop further environmental requirements within them. In some cases, the specific international areas, such as climate change, international trade, illegal, unreported and unregulated fishing, or access to justice in environmental matters already have a significant and well-established international legal framework. In other cases, the areas have not yet seen this level of normative development. Examples include the incorporation of the SDGs into EU policies, the consolidation of mechanisms for the sanctioning of international environmental crimes, the management of migration flows for environmental reasons, and the construction of an enabling normative framework that responds to the emergence of the new paradigm of environmental justice.

These are the central questions that are analysed in this collective book. In this sense, the various chapters offer a cross-cutting analysis that focuses especially on how the EU can strengthen its international position and exert greater legal and political influence, bolstering its regulatory power to promote better international environmental governance. Thus, throughout the book, complex phenomena that are global in scope and require a joint multifaceted approach are examined from a predominantly legal perspective. This does not, however, prevent the authors from also emphasizing the relevance of the implicit scientific, economic, and political issues that determine the context in which the rules are produced, their content, or the values that guide the purpose of EU law. Yet the actions of the EU as a global player have also had a seminal effect, demonstrating a real capacity to transform the analysed sectors and regimes through the strengthening of legal instruments adopted within the EU legal system to implement the environmental rules in these various areas and to promote them at the international level.

Especially in the case of climate change, biodiversity, fisheries, and multilateral trade, the authors highlight the influence of the EU from a international perspective. However, they also put forward a number of proposals to strengthen the EU's leadership. For instance, a cross-cutting issue present in all these areas concerns the scope of the EU's external competences, which requires not only the strengthening of cooperation between the member states and the EU in terms of external representation and the exercise of mixed competences, but also a better and more explicit demarcation in order to avoid overlapping and ensure that the environmental dimension is included and extended to all areas of EU policies – an aspect that is considered essential to reducing pressures on the environment. Another highlight is the need for EU action to facilitate the preservation of a number of major multilateral environmental agreements in order to enable their implementation, at least in part, on the currently established terms. The EU must continue to sustain the ongoing efforts that have been made thus far. This book also points to some practicable opportunities such as the extension of environmental conditionality, for example in the framework of climate change, multilateral trade, or the protection of biodiversity. This can be done by taking account of biodiversity protection requirements in the EU's international cooperation policy, through the tariff preference system, or by incorporating specific chapters on sustainable development in new regional trade agreements. Other scenarios, by contrast, would require a more extensive modification of EU regulations, including a new reading of the founding Treaties that was more open to considering increasingly pervasive phenomena, such as migration for environmental reasons.

The second chapter, which was prepared by Dr. Mar Campins Eritja, highlights the EU's role in setting targets for the reduction of greenhouse gas emissions within international forums since 1990 and points to some of the difficulties that have hindered EU leadership in the climate change regime, not only endogenous and structural difficulties, but also exogenous and conjunctural ones.

Among the former, Campins Eritja focuses on the difficulties that stem from the shared exercise of EU competences. International negotiations on climate change put a spotlight on the model developed by the EU, which is characterized by a discourse around its "responsibility" on the international stage with regard to the protection of common goods. However, the shared exercise of EU competences and the weak coordination among its institutions continue to illustrate the tensions between the supranational and intergovernmental dimensions. Along the same lines, Campins Eritja examines the impact of the application of the principle of integration set out in article 11 of the TFEU in order to point out the multiple limitations that its application suffers from and the little impact that the judicial control of the EU Court of Justice (EUCJ) has had in this specific area in particular.

In relation to the exogenous and conjunctural difficulties, Campins Eritja has chosen to address two aspects that currently affect the EU's role in the international climate change regime in a very significant way. First, she delves briefly into the geopolitical consequences of the position the United States (US) has taken on the Paris Agreement, which force the EU to seek a stronger alliance with China in order to maintain commitments made in the framework of this international agreement. Second, she examines the impact of Brexit on the EU's climate policy, which, in general terms and pending the articulation of a new agreement governing the future relationship between the United Kingdom (UK) and the EU, is considered negative for the credibility and ambition of the EU's internal action and for the effectiveness of its climate diplomacy on the international stage.

The third chapter, which was prepared by Dr. Teresa Fajardo del Castillo, focuses on the relevance of the EU in the international biodiversity regime. This is particularly significant given that 2020, before the arrival of the COVID-19 pandemic, should have been the year that marked a turning point in this international regime.

In this regard, the development of the post-2020 Global Biodiversity Framework of the Convention on Biological Diversity within the United Nations has allowed the EU and its member states to show their clear commitment to leading the Biodiversity Summit that was originally scheduled to be held this year in China. In this context, Fajardo del Castillo critically considers the poor results obtained so far and analyses the extent to which the objectives set out in the EU strategy, far from having been achieved, need to be renewed without delay and include even greater commitments.

From this starting point, Fajardo del Castillo examines the legal instruments through which the EU can strengthen the objectives of the new biodiversity strategy within the framework of its external competence, which is still very much marked by conflicts between the Council, the European Commission, and the member states. Specifically, she highlights the possible consolidation of biodiversity diplomacy, which is currently in the hands of the European External Action Service (EEAS). Although its content is yet to be fully defined, some features have begun to take shape to inform both the EU's internal policies and its external action. It is worth mentioning Fajardo del Castillo's final reflection on the harmonization of the EU Biodiversity Strategy 2030, which should have been adopted in April 2020, and the green multilateralism promoted by United Nations institutions and the Conference of the Parties to the Convention on Biological Diversity. This strategy has been criticized because of the EU's problems in implementing environmental legislation at the national level and because it limits the EU's ambitions to the external level, but, even so, it is valued positively by the author because of its continuity with previous instruments and its ambitious objective.

A highly relevant aspect of our study is the role of the EU in promoting sustainable development by incorporating environmental and sustainability criteria into the mechanisms of multilateral trade in which the EU participates. This is addressed in the fourth chapter, which was prepared by Dr. Xavier Fernández-Pons, who examines the EU's most recent initiatives to achieve this objective.

Fernández-Pons highlights the limited nature of the EU's Generalised System of Preferences and its special incentives for sustainable development and good governance (GSP+), which are applicable to vulnerable developing countries that have committed themselves to implementing various basic international conventions on human and labour rights, the environment, and good governance. Fernández-Pons also analyses the impact of including a special chapter on trade and sustainable development (so-called TSD chapters) in new regional trade agreements established by the EU with third countries. In his analysis of the EU-Mercosur agreement, he points out the holistic vision of these specific chapters through their inclusion of explicit and multiple normative connections between the economic, social, and environmental aspects, but he also warns of the need for their effective practical application in order for them to be truly relevant.

In addition, Fernández-Pons considers the promotion of sustainable public procurement and the way in which the incorporation of criteria linked to sustainable development and social and environmental aspects into EU law can serve as an example for other countries, especially for the parties to the current WTO Government Procurement Agreement (GPA). The chapter closes with an analysis of two important mechanisms. The first is the recent reform of EU anti-dumping regulations, which at least partly entail the consideration of so-called social dumping and ecological dumping, two notions that have traditionally been ignored in the multilateral trading system and that have been met with strong resistance from countries such as China and the Russian Federation. The second concerns various EU trade measures related to carbon footprint, some of which are now being promoted through the European Green Deal. These measures, however, continue to pose major uncertainty and question marks hang over their compatibility with the WTO multilateral regime.

The fifth chapter, which was prepared by Dr. Xavier Pons Rafols, is devoted to a more specific aspect, namely EU action to eradicate illegal, unreported, and unregulated (IUU) fishing. After presenting the global phenomenon of IUU fishing, Pons Rafols points out the limitations of international regulations both in marine areas under the jurisdiction of the member states and on the high seas. Pons Rafols also addresses the international legal conceptualization of the phenomenon.

Pons Rafols highlights the scope and content of EU regulations 1005/2008 and 1010/2009 to prevent, deter, and eliminate IUU fishing in line with the EU's international commitments. He gives particular attention to mechanisms with a market approach, such as catch certification and the strengthened system of control and inspection of third-country fishing vessels by port state authorities. The international effect of these regulations and the use of so-called yellow, green, and red cards to identify non-cooperating third countries is one of the main aspects reflecting the international impact of EU regulations on IUU fishing. At the same time, Pons Rafols recognizes that requiring third states to comply with international rules and assuming powers to monitor their compliance, including measures of a coercive nature, may be presented as a unilateral requirement of the EU, and it does bring a certain extraterritorial component to EU regulations. However, he also points positively to another factor of international importance, which is the EU's action within the specialized agencies of the United Nations (FAO, IMO, and ILO) that manage the implementation of regional and universal multilateral agreements related to fisheries, IUU fishing and the safety of vessels and workers in the fishing sector.

In the sixth chapter, Dr. Alex Peñalver-Cabré analyses the EU's position on access to justice in environmental protection. After reviewing various international instruments, Peñalver-Cabré analyses the EU's participation in the 1998 Aarhus Convention, paying particular attention to the modalities of access, questions of standing, and requirements for the exercise of this right. The EU's participation in the Aarhus Convention has led to the adoption of a series of legislative and non-legislative measures. Peñalver-Cabré examines these measures and how they have been extended. His analysis values the Aarhus Convention's control mechanism for compliance and finds that it could also be extended to other international environmental agreements.

However, Peñalver-Cabré also points out that a certain duality, which mainly affects the material or substantive scope, can be observed in the level of stringency demanded in the implementation and application of these measures, amounting to a kind of double standard that depends on whether they are aimed at the EU itself or its member states. He illustrates this situation through an in-depth analysis of the 2017 Communication from the Aarhus Convention Compliance Committee ACCC/C/2008/32, in which the Committee considers that neither EU legislation nor the case law of the EUCJ complies with the provisions of the Aarhus Convention with regard to access to justice. This situation is clearly reflected in the external dimension of the EU's action in this field, given that it is difficult for the EU to demand a level of compliance from third countries that it does not demand from its own institutions and bodies, as this weakens the EU's role before the bodies that implement the Aarhus Convention.

In other areas, the EU is much less involved and its role on the international stage is consequently less relevant. In the seventh chapter, Dr. Susanna Borràs Pentinat analyses the role of the EU in the face of migratory crises caused by environmental degradation. In this area, the international response, including that of the EU, has been more limited and the protection of people who move for environmental reasons remains insufficient, especially in a context of climate emergency.

In her chapter, Borràs-Pentinat calls for explicit recognition and greater and more effective protection for environmental refugees by the EU. To this end, after examining the various causes of migration in the EU, she focuses on the international legal aspects of environmental migration through the work done by the Office of the United Nations High Commissioner for Refugees (UNHCR) and through the 1951 Convention. She then analyses the legal framework offered in this context by the EU, both in its internal dimension (the management of immigration and refugee flows into EU territory) and internationally (the inclusion of migration and asylum policies in external relations with third countries).

Borràs-Pentinat notes the lack of an appropriate policy response or a structured regulatory framework and identifies a number of constraints, such as the absence of protection for these refugees when moving from one member state to another or against forced return, as well as the creation of a series of obstacles (the securitization of borders, the externalization of border controls, etc.) that fail to take account of the humanitarian urgency. Although the Treaty of European Union (TEU) and the TFEU can provide a sufficiently broad mandate for a revision of the EU's asylum and immigration policy to cover these persons as well, there is no EU instrument or coherent policy to cover them, and their protection is therefore left to the discretion of the member states. Borràs-Pentinat does, however, point to encouraging signs in the midst of this scenario, thanks to the work of EU institutions to add a new dimension to EU policies which would make it possible to address the issue holistically within development strategies and to consider new measures to explore this type of protection and make room for the category of environmental migration in the future.

The severity of the environmental, social, political, and economic impact of the current economic model has led to the need to criminalize various kinds of particularly serious conduct that constitute so-called international environmental crime. This is the issue addressed by Dr. Maria Marques-Banque in the eighth chapter. Marques-Banque focuses her analysis on the use of criminal law and, in particular, on criminal sanctions in the event of non-compliance with obligations established in some multilateral agreements on the environment.

To this end, Marques-Banque first identifies the criminal law provisions in certain multilateral agreements that either directly or indirectly contain such measures, and then examines the limitations arising from developments in EU criminal law and their relationship with international law. Based on these same multilateral agreements, Marques-Banque employs a comparative methodology to undertake an in-depth analysis of the impact of Directive 2008/99/EC on the protection of the environment through criminal law, ultimately highlighting its direct impact on the sanctioning of a breach when related EU environmental legislation exists. By way of illustration, Marques-Banque examines the system of criminal sanctions in two cases of particular importance, namely the illegal trade in tropical timber and illegal, unreported, and unregulated fishing. Despite an initially positive assessment, however, Marques-Banque observes that there is still ample room for the EU to strengthen its international credibility in these areas with regard to compliance with international environmental law and the EU's role as a key political actor in the drafting and negotiation of such legislation.

The ninth chapter, which was prepared by Dr. Mar Aguilera Vaqués, focuses on the incorporation of the SDGs into EU policies. Aguilera Vaqués starts with the enormous efforts that have been made by the EU to promote and implement sustainable development. After offering a historical overview of the evolution of this concept over the course of the United Nations conferences, she addresses the EU's adoption of Agenda 2030 for Sustainable Development in 2015 and examines its Sustainable Development Strategy.

Aguilera Vaqués looks, in particular, at how the EU has sought to reinforce environmental priorities by incorporating them into some EU policies and taking an approach that relies on the establishment of behavioural obligations addressed to member states, since they bear the primary responsibility for implementing Agenda 2030. However, as Aguilera Vaqués points out, these objectives are clearly far from being achieved, and the EU's efforts have so far yielded very limited results, clearly weakening the EU's role as a model for the rest of the world.

Not only does the EU need to strengthen its leadership on the SDGs, but it should also limit the negative impacts of its policies, which affect the achievement of the SDGs by third states. This requires the EU to ensure a horizontal approach to the SDGs in its policies and to fully integrate them into its policies by providing clear guidance to both its institutions and member states. In this context, the 2019 European Green Deal, which Aguilera Vaqués addresses in the last part of the chapter, should provide a clear response that can decisively bring about deeper and more structural transformations.

In the tenth and final chapter, Dr. Jordi Jaria-Manzano examines the classic assumptions underlying the concept of sustainable development before turning his attention to the new paradigm of environmental justice as a matrix for interpreting the global environmental crisis. Jaria-Manzano questions the concept of sustainable development as a mechanism to modify the current institutional structure and the regulatory frameworks that drive the hegemony of the current model. By contrast, the notion of environmental justice takes into account the scarcity and vulnerability of natural resources, the social distribution of environmental harms and benefits, and the diversity of human communities in such a way that it becomes possible, according to the author, to confront relations of unequal exchange in their entirety. From this perspective, the new paradigm of environmental justice allows for a critical approach to current legal solutions and addresses the exploitation of natural resources within an institutional and legal framework that essentially seeks to reduce inequality and injustice. According to Jaria-Manzano, it takes on a global dimension and demands particular responsibility from institutional actors and especially from the EU.

However, Jaria-Manzano is pessimistic about the incorporation and development of this new paradigm in the EU model, which is strongly linked to the hegemonic idea of sustainable development. For this reason, he does not predict significant changes through conventional decision-making processes in the EU. Nevertheless, he argues that environmental justice, insofar as it is linked to the adequate management of scarce finite resources in a fair manner, is closely related to the idea of sustainability and social justice, which form part of the axiological foundations of EU law. Accordingly, he contends that the fundamental values contained in EU law concerning the protection of human rights would permit a reading in line with a progressive transformation of the rights paradigm, in accordance with a new idea of well-being based on the concept of quality of life and linked to responsibility and participation.

Along these lines, the book shows that the EU must continue to make progress in the more cross-cutting areas of sustainable public procurement and measures against so-called ecological dumping, in strengthening the obligations of port states and flag states in the area of fisheries, in actions that affect EU criminal law and have an impact on compliance with international environmental law, or in considering new ways of protecting and accommodating environmental migration. Various chapters also express the need for a greater EU presence, at least in those areas where there is a clearer international consensus on the necessity of coordinated and global action. They also stress that the EU must strengthen its participation in United Nations initiatives, where the EU must act promptly to inspire a better understanding of environmental problems and their consequences in a more equitable way.

In short, the EU – now more than ever – has to assume the role of champion for green multilateralism and to make the most out of the many strategies which it has already adopted in the various analysed sectors and which have been designed precisely to export the EU's regulatory model and the stringency of its requirements at the international level. Clearly, the EU needs to take action. Indeed, this is the underlying objective of the book and the contributions made by each author. With every passing day, there is an increasingly urgent need for international leadership which is both shared and multilateral, leadership in which the EU finds its role as an international actor and drives home the value of its environmental policy, despite some of its own endogenous weaknesses.

There is no doubt that the EU has contributed very significantly to the emergence, development, and consolidation of international environmental law and to the construction of the various environmental regimes as we know them today. In recent years, however, the EU's leadership has been compromised, if not undermined, by an excessive fragmentation of its own institutional and regulatory framework, and this has affected its international impact. In addition, the financial and economic crisis of 2008 was a very important turning point, which clearly showed the vulnerability and limitations of the EU's supposed leadership in an increasingly multipolar international context. Since then, the EU has tried to overcome the impasse by taking the lead in different initiatives pursued within the framework of various environmental regimes or regulations, with the aim not only of presenting a strict and rigorous EU environmental policy to the entire international community, but also of regaining its international credibility in the construction of a new model of international environmental governance.

On the other hand, the book is a result from the activities of the Jean Monnet Chair in EU Environmental Law which, thanks to funding from the European Commission within the framework of the Erasmus+ programme, have been pursued between 2017 and 2020 by a group of researchers and professors in public international law, constitutional law, criminal law, and administrative law at a host of Spanish universities. In a complementary manner, they share a common interest: legal protection of the environment. Together, they pursue this interest from the viewpoints of their respective legal disciplines, especially in relation to the instruments offered by EU law. Therefore, from a methodological viewpoint, the methods of legal science have prevailed in the preparation of each chapter, reflecting the legal specialization of each author. For this reason, each analysis brings together the various relevant elements of the international and EU legal systems, adopting a dual perspective that addresses not only the rules and institutions, but also the processes for their formulation and execution within the framework of each area. It has therefore been possible to identify various issues that the EU's participation raises and to assess the effectiveness of its rules as a response to the issues within the international regulatory frameworks in question.

In this sense, it is necessary to warn that the drafting of the chapters of this collective work was completed in April 2020. It is well known that the dynamics of the subject-matter we are presenting is very important. We are therefore aware that there may be a certain time lag between the completion of our study and its final publication. However, we believe that as a collective work, it offers an approach that allows us to raise the major legal issues of the EU's participation in the international arena, and the need for its participation in multilateral cooperation mechanisms, beyond the specific timetable and the very rapid development of current events.

As a final comment, it should be noted that EU action in the various analysed sectors has been greatly affected by the events of spring 2020. The EU must now face up to the consequences of the COVID-19 pandemic, a challenge that is unique in its scope and size, and that has its origins, like most epidemics in human history, in a change in the relationship between humans and biodiversity. The repercussions of this global crisis on the political agenda of the environmental negotiations in which the EU is involved have yet to be felt. While they may be difficult to measure at present, they will certainly be very significant.

The economic crisis that follows the health crisis will undoubtedly impact the EU's standing on the international stage. It will also call into question many of the achievements that have been taken for granted within the framework of the sectors and international regimes addressed in this book. To begin with, even before they have been put into effect, the EU will have to readjust its strategies in the various environmental sectors in question. In addition, the brand-new European Green Deal, which has been designed as a roadmap for a period of economic recovery and growth, will now have to be rethought. It is also true, however, that some of the initiatives that it contains, such as those relating to decarbonization of the economy, increasing the use of renewable energy sources, and modernizing the transport sector, are now even more relevant, because they were designed to promote growth that is more in line with the demands of the climate emergency and biodiversity.

For the time being, the health crisis has put a spotlight on the need for greater cooperation at the EU and international levels to address its causes and prevent a spill over of any social, political, and economic consequences that it may have in the medium and long term. The international community, the EU, and EU member states are already developing plans to overcome the crisis. Many actors are calling for these plans to be geared towards accelerating the fight against climate change and biodiversity loss. The EU's current proposals already incorporate some elements that move partly in this direction.

The overly anthropocentric focus of the sustainable development paradigm and the perpetuation of economistic perspectives seem to have led us, finally, to a point of no return. The "new normal" that we enter after the current global health crisis – which at present raises major uncertainties about the evolution of our development model and of our societies as we now know them – will also require new forms of international environmental governance and greater international cooperation, in which the EU must not only take part, but must lead.

Notes

1 Decision No 1386/2013/EU of the European Council and of the Council, of 20 November 2013, on a General Union Environment Action Programme to 2020 "Living well, within the limits of our planet", OJEU L354, p. 199.
2 *Ibid.*

2 Some challenges for strengthening the EU's international influence in the climate change regime

Mar Campins Eritja

1. Introductory remarks

Since the beginning of climate change negotiations in the late 1980s, the European Union (EU) has sought to take the lead in building this international regime, and there is no doubt that it exerted a major influence in the first stages of this process.[1] The EU exercised this leadership vis-à-vis its member states by adopting measures such as the implementation of the EU Emissions Trading Scheme (EU ETS) in 2005[2] and, at a more global level, having been able to rescue the Kyoto Protocol to the United Nations Framework Convention on Climate Change[3] (Kyoto Protocol) after the United States' rejection in 2001. The EU also reached an agreement with the Russian Federation in 2004, which allowed the Kyoto Protocol to come into force in 2005. In those years, the EU set the pace and the rest, more or less, followed.[4] Since 2005, the EU has focused on meeting its Kyoto targets and drawing up a new international agreement for the following period, despite the scenario of a world economic and financial recession, and the detachment of some member states.

Alongside the United States (US) and China, the EU is considered today as one of the key players in climate change negotiations, but its role had progressively diminished after the Poznan and Copenhagen summits in 2008 and 2009.[5] At the 15th Conference of the Parties (CoP) in Copenhagen, the EU lost momentum by failing to secure support for a treaty with binding targets and timetables, and ended the summit with a sense of frustration and (some might say) political humiliation. From 2011, the new negotiating mandate that was to lead to the Paris Agreement of 12 December 2015[6] allowed the EU to regain, at least temporarily, its status as the most reputable leader; in Paris, the EU had the support of the US, still under the Obama administration, and China's President Xi Jinping.

Part of this upturn was due to the fact that the EU approached the Paris negotiations with a greater dose of flexibility and realism, especially with regard to the design of the agreement. On the one hand, it favoured an institutional architecture far removed from the strict binding emission reduction targets and timetables it had previously advocated for in Copenhagen. On the other, its position in the negotiations continued to revolve around the example it set in implementing its climate policy, which it sought to escalate to the global level by supporting a rule-of-law system and the inclusion of its targets and binding greenhouse gas (GHG) emission reduction policies. The EU encouraged this strategy through its active participation in the *High Ambition Coalition* (HAC), intensifying dialogue and continuous interaction with third states and promoting the advance of international standards based on the previous progress made by the parties that could lead the process.[7] This made it possible to isolate potential vetoes to the Paris Agreement and gather the necessary support to achieve its adoption.

With this strategy, which it has continued to pursue through its ongoing action on climate diplomacy,[8] the EU achieved a hybrid agreement in 2015, with a system of bottom-up emission reductions through Nationally Determined Contributions (NDC) combined with rigorous international review. Unlike other instruments, the Paris Agreement is based on the idea that it is easier for parties to make unilateral and voluntary commitments than to agree on multilateral commitments with mandatory reduction targets. The Paris Agreement also applies a managerial approach based on two main elements: on the one hand, the clear articulation of a global interest (to limit the increase in the global temperature of the planet to within 2°C of pre-industrial levels, and to make efforts to limit it to 1.5°C, as well as to increase the capacity of the parties to adapt to the adverse impacts of climate change and to move towards a low GHG emission model and climate-resilient development); and, on the other, the commitment to transparency with regard to the progression and ambition of the mitigation and adaptation targets (parties are required to communicate their increasingly ambitious NDC every five years, inside an overall time frame from 2020 to 2030).

In this context, this chapter aims to analyse whether the EU truly represents the normative power that it aspires to become within the climate change regime, and to identify some obstacles to achieving this objective. The first two sections focus on two important endogenous structural developments that impact the role of the EU in the international arena. Section 2 discusses the difficulties involved in the shared exercise of EU competences. International negotiations on climate change highlight the model developed by the EU, characterized by the discourse around its "responsibility" with regard to the protection of common goods. However, the shared exercise of EU competences and the weak coordination between institutions still reflects the tension between the supranational and intergovernmental dimensions. Section 3 focuses on the application of the integration principle into the external EU climate change policy. Announced in article 11 of the Treaty on the Functioning of the European Union (TFEU), it gives the EU a comprehensive tool to provide sound and appropriate answers to complex problems such as global warming. Nevertheless, as the EU measures relate to a wide range of areas, this might lead to conflict between different legal regimes. Additionally, although EU legislative acts are addressed to its member states, some measures clearly affect non-member states as well, raising the issue of the extraterritoriality of the EU's legal activity.[9] Sections 4 and 5 analyse two temporary exogenous developments that may affect the EU's external climate change action in the future. Section 4 is devoted to how the climate change policy of the US administration has introduced considerable uncertainty in the United Nations Framework Climate

Change Convention (UNFCCC)[10] regime and stresses the need for the EU to create a stronger alliance with China in order to implement the Paris Agreement, and section 5 focuses on the impact of Brexit on EU climate policy and the challenges raised and still to be resolved. The chapter concludes with some final remarks and proposals.

2. The shared exercise of EU competences

The EU and its member states are all parties to the UNFCCC, the Kyoto Protocol, and the Paris Agreement, and have participated jointly in their negotiations. Following established practice with respect to other multilateral environmental agreements, article 22 of the UNFCCC, article 24 of the Kyoto Protocol, and article 20 of the Paris Agreement allow "regional economic integration organizations" to become parties, while requiring the EU to declare the extent of its competence in its instrument of ratification. Since then, the EU has sought to lead efforts to adopt specific targets for the reduction of GHG emissions and to design new mechanisms for this purpose, while promoting the incorporation of certain principles into this international regime.

The nature and specificity of the EU as an international organization and the lack of a consolidated institutional structure to facilitate the exercise of international competences, despite the efforts of the European External Action Service (EEAS),[11] affect its capacity as an external regulatory power in the fight against climate change.[12] Within this international framework, both the regulatory and operational problems that the EU has to face are conveyed and its role as provider of political objectives and legal techniques is made more difficult. Jordan et al. stress how, in a context of dispersal of international leadership, the complexity of the EU's decision-making system, the progressive weakening of its governance, and the growing divergence between member states are a real impediment.[13]

The external action of the EU has never been problem-free and it has traditionally required an additional effort to project its regulatory power. This issue, which is of a constitutional nature and significance, is mainly linked to the regime of competences and the legal basis for EU action. Both elements limit the EU's scope for action in international negotiations and highlight the tensions between its supranational and intergovernmental dimensions.[14]

Article 5 of the Treaty on European Union (TEU) states that the EU can only act within the powers conferred upon it by the Treaty. Since the early 1990s, the EU's participation in the international climate change regime has been based on the existence of a shared competence between the Union and its member states, as set out in article 4 of the TFEU.[15] Its exercise by the EU is therefore subject to the application of the principles of proportionality and subsidiarity,[16] while its member states retain the capacity to perform regulatory or innovative functions in the legal system, and have the power to adopt their own legislative, administrative, or regulatory actions.

From the perspective of the EU's external relations, the Treaty requires the existence of substantial competences and the choice of an appropriate legal basis.

Article 191 of the TFEU includes, among the EU's competences, the promotion of climate change mitigation at international level. It therefore follows that the European Commission has an obligation to formulate a policy in this area within the powers conferred by the Treaty. However, this same provision states that the EU shall only *"contribute"* to the preservation and improvement of the quality of the environment. In addition, article 193 of the TFEU expressly reserves the right of member states to maintain or adopt measures of enhanced environmental protection.

The TEU also establishes an explicit link between sustainable development strategies and the EU's external relations; thus, article 3.5 of the TEU specifies that, in its relations with the rest of the world, the EU "shall contribute to . . . the sustainable development of the Earth . . . as well as to the strict observance and development of international law" as one of its objectives. Furthermore, according to article 21.2 of the TEU, the EU

> shall define and pursue common policies and actions, and shall work for a high degree of cooperation in all fields of international relations, in order to: . . . help develop international measures to preserve and improve the quality of the environment and the sustainable management of global natural resources, in order to ensure sustainable development.

This means that the fulfilment of the obligations arising from international agreements on climate change has a dual perspective. On the one hand, the EU takes as its starting point the sharing of responsibilities between the member states for the mitigation of GHG emissions and the implementation of adaptation actions. On the other, it adopts a common sectorial approach that allows it to adapt to the different needs that arise in the various fields of action.

Besides, article 216 of the TFEU expressly recognizes the EU's treaty-making power, a recognition that is merely the legal embodiment of a practice that the EU had been carrying out informally since the late 1990s.[17] Furthermore, the EU Court of Justice (EUCJ) warned in its Opinion 2/2000 on the legal basis for concluding the Cartagena Protocol[18] that international agreements may also have as a legal basis in article 192.1 of the TFEU when they are not simply related to development cooperation (in this case, specifically provided for in article 191.4 of the TFEU). In its 2011 Decision on the application of the EU ETS Directive to the civil aviation sector,[19] it ruled that the mere confirmation of a future international action was sufficient to recognize the EU's right to act and that the EU may intervene at international level when the required action has not finally been passed (in this case, the adoption of an international agreement within the International Civil Aviation Organization [ICAO]).

Insofar as this is an internally shared competence, at an external level it translates into the need to pass mixed agreements, the negotiation process for which is regulated indirectly by article 218 of the TFEU (mixed agreements are not expressly provided for in the Treaty). While the EU tries to show a certain unity on the international stage, this is often an obstacle to the EU's capacity to pursue a clear and coherent policy in any particular field, and may occasionally prevent the EU from fully deploying its regulatory power due to the requirement that it be unanimously adopted by the member states.

Although the Lisbon Treaty and the launching of the EEAS have thus clarified the procedure for concluding such agreements, it is sometimes quite difficult to know in advance and for sure what form the EU's external action will take in every situation. Even so, the role of the EEAS is central to the EU's coherence, not only in supporting the European Commission in the implementation of its mandates, but also in ensuring consistency of action when the EU faces major cross-cutting challenges, such as the fight against climate change.[20]

However, there are still major hurdles in this rationalization exercise. On the one hand, the international representation of the EU remains the monopoly of the European Commission. This has been questioned by the member states from different angles, as they consider that the external representation cannot be used by the European Commission to set the extent of a shared competence. On the other hand, a number of actors continue to be involved in the exercise of external representation. The EU delegation may interfere with member states' representatives in their bilateral or multilateral contacts and vice versa, creating further confusion for third states.[21] This situation is made even more serious by the fact that the exercise of such power is not framed by precise rules defining the distribution of their respective international responsibilities beyond the duty of loyal cooperation set out in article 4.3 of the TEU.

Besides this, it is worth noting that, throughout the international negotiations, member states must agree on common EU positions within the Council. The need to agree on a new common position with regard to proposals from other international actors is often a source of frustration, with endless coordination meetings due to the inflexibility of the Council's mandates. This situation of uncertainty is exacerbated by the fact that the TFEU grants important concessions to member states and requires their unanimity on some areas related to climate policy, which can block the EU action.[22] This requirement of unanimity, included in article 194 of the TFEU on EU energy policy, hampers the adoption of legislative measures and explains why it is so difficult for the EU to address the energy sector in an efficient manner, as this is a strategic sector in the fight against climate change in which the EU has only very limited competence. With regard to these institutional obstacles, the EU needs to have more explicit competences on climate change and energy policy, and to define the overlaps between climate change regulation (especially with regard to the EU ETS) and energy regulation (especially with regard to the promotion and use of renewable energies).[23]

3. A comprehensive and integrated approach to fight climate change

Another important provision of the TFEU is article 11, according to which "Environmental protection requirements must be integrated into the definition and implementation of the Union's policies and activities, in particular with a view to promoting sustainable development". The principle of integration, which was linked from a very early stage to EU environmental policy,[24] also has an external dimension. This clearly arises from articles 3.5 and 21.2 of the TEU, which set out the EU's commitment to contribute to sustainable development and the preservation of the environment in its international action. This obligation is further reinforced by the requirement for horizontal coherence, linked to the interinstitutional mechanisms set out in article 7 of the TFEU, according to which "The Union shall ensure consistency between its policies and activities, taking all of its objectives into account".

The principle of integration provides the EU with a tool to give consistent answers to complex problems such as the one we are dealing with now. In short, it is a question of strengthening the coordination of EU climate policy from a cross-cutting perspective. The European Commission has, at least in part and from a formal point of view, welcomed this approach in several policy documents. Thus, after the financial collapse of 2008, the EU adopted its "2020 Strategy" in 2010, which set out certain guidelines for achieving greener growth through a roadmap with emission reduction targets as part of long-term sustainable growth.[25] Likewise, the principle of integration is implicit in the concept of the circular economy promoted by the European Commission in 2015 as a model of production and consumption.[26] A few years later, this approach has been extended (at least on paper) to the whole of the economic system through the "Roadmap 2050: towards a carbon-free economy",[27] which is to be developed through regulatory review in sectors such as the emissions market, transport, renewable energies, and energy efficiency,[28] and should allow enhanced sustainability and the transition to a safe, neutral, and climate-resilient economy, which is key to ensuring the long-term competitiveness of the EU economy as a whole.[29] The European Commission's central tool for developing the objectives of this circular and carbon-neutral model in the period 2019–2024 will be the recently adopted European Green Deal.[30] The president of the European Commission, Ms. von der Leyen, has insisted on the need for this integral transformative approach in order to redesign, for example, policies for the supply of clean energy to the entire economic system, including industry, production and consumption, large infrastructures, transport, food production and agriculture, and construction.[31]

In particular, article 11 of the TFEU authorizes the EU to adopt policy measures in areas outside the environmental field that may affect the objectives of article 191 of the TFEU, whether its competence is shared or exclusive. An example of this is precisely the contribution of certain measures to combat climate change adopted in EU policies that do not have an environmental dimension.[32] As De Sadeleer points out, environmental protection measures cannot be limited to the sectors traditionally covered by environmental policy, but must be decompartmentalized and extended to all EU action.[33] In fact, the TFEU does not contain a single legal basis for the adoption of these measures and allows recourse to other legal bases in addition to environmental policy. This should make it easier for all sectors and activities to contribute to reducing GHG emissions and to adapt to climate change through a holistic approach that integrates this requirement in these areas of action.

However, there still are significant limitations to the application of this principle.[34] It is difficult to put into practice and to verify, and judicial control of compliance has had little significant impact in this field.

When applied internally by the EU, the general wording of article 11 of the TFEU does not reveal the exact scope of the obligation contained in this provision. The provision clearly states that environmental requirements "must be integrated into the definition and implementation of the Union's policies and activities", which suggests a strict interpretation of the principle that goes further than the mere requirement to assess the compliance of EU measures with environmental protection criteria.[35] However, it is difficult to determine to what extent the margin of discretion of the European Commission is actually limited when presenting its proposals to the European Parliament and the Council. In this regard, the impact assessment reports of the environmental, economic, and social effects of the European Commission's proposals (which remain the best way of respecting the principle of integration), introduced in 2003,[36] and which have since been prepared with respect to legislative initiatives, have not resulted in a greater contribution to sustainable development. In general terms, as far as the environmental dimension is concerned, it has been limited to the comparison of environmental aspects between various policy options in the process of preparing legislative acts.[37]

In the external dimension, and beyond the fact that international agreements are also subject to this environmental impact assessment, the reference in article 191.1 of the TFEU to "measures at international level to deal with regional or worldwide environmental problems" also raises the issue of the application of the principle of integration outside the territorial scope of the EU. In this case, however, the action of the EU is – like that of its member states – subject to the international law concerning the extraterritorial application of domestic rules.[38]

The EU's actions to combat climate change may clearly have an impact on economic operators in third countries, just as they affect various areas of action across the board. The application of its measures to the civil aviation sector, for example, has raised major conflicts that have affected its leadership in this area. Directive 2003/87/EC on the greenhouse gas emission allowance trading system[39] has traditionally been the normative pillar of the fight against climate change within the EU, setting targets for the reduction of GHG emissions. Thus, the economic sectors covered by the EU ETS are currently committed to a reduction of GHG emissions of -40% below 1990 levels by 2030. The Directive was amended in 2008[40] with the aim of including aviation activities in the EU ETS, covering non-EU aircraft landing at or departing from EU airports. A number of third-country airlines challenged its consistency with the principles of international customary law, and the provisions of international treaties, by way of a preliminary ruling from the High Court of Justice of England and Wales.[41] Although the EUCJ upheld the validity of the EU legislation in 2011, the EU was forced to adopt a temporary derogation from the suspension of penalty measures for third-country aircraft operators that failed to report emissions and were obliged to surrender allowances for 2012 (Council Decision 377/2013/EU).[42] Subsequently, Regulation (EU) 421/2014[43] allowed for an extension of the temporary derogation and excluded flights by non-commercial operators emitting less than 1,000 tonnes of CO_2 per year from the EU ETS until 31 December 2020. Regulation (EU) 2017/2392 again extended this derogation until 31 December 2023.[44]

In short, the question of the practical and effective insertion of the principle of integration remains inadequately resolved at the EU level. The challenge is how to strengthen the use of the tool of EU policies in the pursuit of the aims of sustainable development, the fight against climate change, and the decarbonization of the economy under the terms of the Paris Agreement.[45]

4. The United States' withdrawal from the Paris Agreement and the EU's stronger alliance with China

Over the years, the EU's climate change action has sought consistency between its internal dimension and regulatory development at an international level, in which it strives to serve as a model as far as objectives, principles, and mechanisms are concerned. In doing so, the EU aims to reaffirm the projection of its influence beyond its borders and has therefore stressed the importance of global action in the fight against climate change. To do so, it needs to permanently push third states in this direction – especially those with more developed economies – by continuously demonstrating that effective measures to reduce GHG emissions are possible.

This interest, beyond any altruistic motivations, has in fact been determined by the need to address the economic and financial crisis that has afflicted the EU since 2008, by guaranteeing, through the involvement of major economic competitors such as the US and China, that the adoption of such measures would not lead to an international competitive disadvantage for the economies of EU member states. Internally, the EU's dependence on energy imports from third countries is another important incentive for building the consensus needed to promote structural reforms and to ensure the security of the energy supply.[46] However, these objectives, which are geostrategic in nature and go beyond strict environmental concerns, have also meant that the EU has had to continually adjust its strategy in relation to other major global players in order to avoid isolation.

In this regard, the election of Donald Trump as the US president in 2016 introduced great uncertainty regarding the development of the Paris Agreement and the UNFCCC itself. Although the official communication of 4 November 2019 to the UN secretary general[47] leaves the door open to the US remaining in the Paris Agreement, its withdrawal will most likely become a reality on 4 November 2020. The US will thus be absent from the Paris Agreement in the future, as it was in the past with regard to the Kyoto Protocol, unless a new administration emerges in the next presidential elections that wishes to return to the Agreement. In this scenario, it would also have to be seen whether Senate ratification would finally be required; this had not been the understanding of the Obama administration, which ratified the Paris Agreement in 2015. If Senate ratification is necessary, the project would be very unlikely to prosper given the clear opposition of the Republican senators.

Until then, the US continues to be a party in the negotiations for the implementation of the Paris Agreement and, in particular, remains bound by the obligations set out in articles 3 and 4 thereof concerning compliance with and maintenance of the NDCs and the level of ambition. Nothing in the Paris Agreement allows the extent of these obligations to be qualified or these commitments to be diluted. Despite this, even before its withdrawal becomes effective, the US position has already resulted in a lack of support from the federal government for measures to mitigate and adapt to climate change, especially those relating to the funding of the Green Climate Fund.

Nevertheless, it should be noted that while in the EU the cross-border dimension of climate change issues has allowed a certain "continentalization" of policy measures, favoured by their supra-governmental nature, in the US the limitations of the federal framework have resulted in a greater voice for certain states, which have taken the lead in the fight against climate change. Through the establishment of coordination mechanisms,[48] these states have become more prominent in the implementation of climate change policies. Although they still have a limited role at the international level and are unlikely to be able to mitigate the uncertainty created by the Trump administration, American coalitions such as We are Still In, US Climate Alliance, Global Covenant of Mayors for Climate & Energy, and America's Pledge, which are present at the CoP and more active than the official national delegation, have been a source of some relief to all the parties to the Paris Agreement.

In this scenario, which is a challenge but also a great diplomatic opportunity, the EU has been forced to strengthen its alliance with China, the world's largest energy consumer and GHG emitter,[49] in order to preserve the stability of the coalition that made the Paris Agreement possible and to allow, at least in part, its implementation under the terms currently settled.

The EU-China energy dialogue was launched in 1994, and recorded a major step forward[50] with the Joint Declaration and Partnership on Climate Change of 2005[51] and the two Memoranda of Understanding between the EU and China of 2006 and 2009.[52] Through these instruments, the EU sought to use its capacity to influence the development of China's climate policy. Although it eventually failed to achieve concrete results at an international level, as evidenced at the Copenhagen Summit in 2009, it allowed for the establishment of other institutional arrangements between the two blocs.[53] It also contributed, for example, to the dissemination in China of the carbon capture and storage technologies covered by Directive 2009/31/EC of 23 April 2009 on the geological storage of carbon dioxide, and to the development of the GHG emissions market, in a way modelled on Directive 2003/87/EC.[54] Besides, China's economic development in recent years has led to negative externalities and significant environmental degradation, requiring a reconsideration of the country's foreign policy in terms of its climate agenda. As a result, the EU-China Joint Statement on Climate Change of June 2015[55] reiterated the collaboration between the two blocs and confirmed the continuation of existing bilateral cooperation programmes, as well as encouraging the announcement of new Chinese NDCs.[56]

Since then, relations between China and the EU in this area have generated a "new" climate of diplomacy (more inclusive but less rule-of-law-based),[57] conditioned by the phenomenon of globalization and by China's geopolitical emergence. The need for new effective action in the fight against climate change formed the basis of the EU-China Leaders' Statement on Climate Change and Clean Energy adopted in Beijing in July 2018[58] and was also an important factor at the last EU-China Summit on 9 April 2019 in Brussels,[59] which served to reconfirm the Chinese commitment to the Paris Agreement.

The next EU-China summit was originally scheduled for September 2020 to be held under the German presidency in Leipzig. However, it has been postponed until the end of 2020 due to the consequences of the COVID-19 crisis. If it is finally convened,[60] it is worth noting that the summit should take place after the US presidential election and before the CoP of the Biodiversity Convention in Kunming, rescheduled for May 2021, and the next CoP of the Paris Agreement in Glasgow, also moved to November 2021. For China, these talks will coincide with the completion of its five-year economic plan and the definition of its economic framework for the period 2021–2025 and should also include some trade-offs beyond the specific mandate on climate change, favouring a broader agreement.[61] This is an optimistic scenario, but the negotiations will face other major challenges and uncertainties: among them, the development of the Chinese mega-project of the Belt and Road Initiative;[62] the negotiation of the EU-China Comprehensive Agreement on Investment,[63] which will undoubtedly be an important platform for discussing new possibilities for collaboration in the fight against climate change as well; and the need to negotiate arrangements related to technology transfer, an area in which there are still strong divergences between China and the EU. So, this might have been a unique though extremely complex opportunity to strengthen this bilateral cooperation.

5. The Brexit and its impact on EU climate policy

The Agreement on the Withdrawal of the United Kingdom of Great Britain and Northern Ireland from the European Union and the European[64] Atomic Energy Community and the resulting British European Union (Withdrawal Agreement) Act 2020[65] open up a completely new scenario for the EU. On 31 January 2020, the United Kingdom (UK) became, in accordance with article 50.3 of the TEU, the first member state to withdraw from the EU.[66] On this date, the transition period provided for in article 127 of the Withdrawal Agreement was opened. This period extends the UK's link with the EU until 31 December 2020, during which time the two parties will have to negotiate their future relationship, which is still very uncertain at present.

Brexit presents both challenges and opportunities for climate change policy in the UK. However, its impact on EU climate policy is generally negative in terms of the credibility and ambition of the EU's internal action and with regard to the effectiveness of its climate diplomacy on the international stage, since the UK has traditionally been a major player in EU climate policy.

In its internal dimension, the UK has been a strong supporter of EU action on climate change over the last 15 years. It was the first member state to set a "net zero" GHG emissions target,[67] and it has supported a policy discourse focusing on technology neutrality for climate mitigation. Its energy market is now fully integrated with that of the EU member states, participating very actively in the EU ETS.

From 31 December 2020, the UK will be regarded for all intents and purposes as a third state. Therefore, from then on, the EU rules governing the EU ETS and the sectors outside it (energy, transport, construction and buildings, waste, and agriculture) under Regulation (EU) 2018/842,[68] will no longer apply to its territory. However, neither Directive 2003/87/EC nor Regulation (EU) 2018/842 refers to Brexit and, in fact, although the Regulation was adopted as early as 2018, the UK is still allocated an overall reduction of emissions of -37% from 2005 levels.

Article 96 of the Withdrawal Agreement only refers to the obligations to monitor and report GHG emissions included in Directive 2003/87/EC, applicable to the UK only until the end of the transitional period, while limiting the UK's commitment to the national GHG inventory and emissions registry systems.[69] Furthermore, the obligation to monitor and report CO_2 emissions from vehicles[70] will only apply until the same date. More generally, the agreement to limit GHG emissions for the period 2013–2020 (Sharing Effort Decision of 2009),[71] as well as the obligations concerning the establishment and administration of emissions registries in the UK, are also terminated at the end of the transition period.

As it stands right now, this would mean that after the transition period, the UK's GHG-emitting facilities would no longer be subject to EU climate legislation; that is, no further auctioning of GHG allowances by the UK or free allocation from the UK would be possible under the EU ETS, nor any trading in GHG emission credits administered by UK operators, and a derogation for flights to and from airports in countries outside the European Economic Area would also apply.[72] The verification system and the EU ETS Union Registry would also be affected,[73] although the EU has already amended the provisions relating to the emissions market to prevent UK emitters from reselling their allowances *en masse* to other member states.[74]

On the other hand, the UK's decision to leave the EU will also have significant implications for the future of the EU's climate change policy in its international dimension, as the UK's exit represents the loss of a major advocate of ambitious and economically sound climate action at global level. The UK, with a permanent seat on the United Nations Security Council and a large and strong network of diplomatic relations, has traditionally been the best platform for the international promotion of the reduction of GHG emissions and the need for adaptation to this phenomenon.[75]

Brexit does not affect the architecture of the Paris Agreement, since it was adopted as a mixed agreement like the UNFCCC and the Kyoto Protocol, and indeed its multilateral nature serves as a safeguard against political disruption in any state.[76] Those are agreements adopted and ratified by the EU and by each of its member states in accordance with their own internal constitutional procedures, which transfer to the international arena the shared competence that they exercise in this area. In this regard, it should be recalled that article 129 of the Withdrawal Agreement clearly states that "during the transition period, the United Kingdom shall be bound by the obligations stemming from the international agreements concluded by the Union, by member states acting on its behalf, or by the Union and its member states acting jointly" and that "In accordance with the principle of sincere cooperation, the United Kingdom shall refrain, during the transition period, from any action or initiative which is likely to be prejudicial to the Union's interests".

The legal implications of Brexit may be more complex with regard to the settlement of the EU's NDCs. The UK may submit a new NDC in accordance with article 4.11 of the Paris Agreement, which allows that "A Party may at any time adjust its existing nationally determined contribution with a view to enhancing its level of ambition". In principle, this would not pose a major problem for the UK, since the Climate Change Act of 2008 already provided for a target of a 57% reduction in GHG emissions by 2032 and 100% by 2050 compared to 1990 levels.[77] In contrast, the joint achievement of the NDCs to which the EU committed itself in 2015 raises major concerns,[78] especially if the EU ETS, as we said, remains the cornerstone of EU climate policy. In this respect, the Grantham Research Institute on Climate Change and the Environment[79] has warned that it will be more difficult for the EU to achieve its NDC target without the UK, as the UK has reduced its emissions at a faster rate than the EU itself.[80] Current projections already also indicate that GHG emissions in an EU with 28 member states, including the UK, are likely to exceed its 2030 target by between five and ten percentage points. The EU would need to at least double its annual rate of reduction, especially in the sectors not covered by the EU ETS, in order to meet its 2030 target. Without the UK, the most plausible scenario is that this situation will worsen, and that the remaining member states and the EU as a whole will need to raise the ambition of their respective NDCs.

Furthermore, the vacuum left by the UK may weaken the European Commission's position vis-à-vis third states and dilute its negotiating capacity, as the UK has also been one of the few member states with significant external action of its own.[81] The UK's absence will limit the EU's diplomatic solvency to jointly negotiate these climate agreements in the future and, outside the EU, the UK may become a "middle power" in climate matters, like Canada within the Umbrella Group or South Korea in the Environmental Integrity Group.[82] The question of which negotiating group inside the climate CoP the UK will finally join will also have important legal implications.

All these internal and external challenges will be on the negotiating table for the new agreement that is to govern future relations between the EU and the UK. The Political Declaration setting out the framework for the future relationship between the European Union and the United Kingdom[83] expressly refers to some of these aspects. Thus, it is mentioned that

> The Parties will retain their autonomy and the ability to regulate economic activity according to the levels of protection each deems appropriate in order

to achieve legitimate public policy objectives such as . . . the environment including climate change. . . . The economic partnership will recognise that sustainable development is an overarching objective of the Parties.

(par. 18)

and that "The Parties should consider cooperation on carbon pricing by linking a United Kingdom national greenhouse gas emissions trading system with the Union's Emissions Trading System" (par. 70); or that "the Parties should cooperate in international fora, such as the G7 and the G20, where it is in their mutual interest, including in the areas of: a) climate change; b) sustainable development; c) cross-border pollution" (par. 75). But the use of the conditional tense does not augur well and, so far, neither the EU nor the UK has provided a satisfactorily clear picture of the future to ensure a sufficient degree of legal certainty for economic operators.

The latest version of the Draft Agreement on the new partnership with the UK, prepared by the European Commission's Task Force in February 2020[84] and supported by the Council's negotiating directives, sets out some of the principles of the climate change negotiation which fits into the overall agreement. Thus, in addition to the commitment with the Paris Agreement and multilateral action (articles 5, 41, and 42), it includes the principle of climate neutrality and nonregression of the level of climate protection, as well as the requirement to seek higher levels of protection (articles 30, 31, 34, and 36). Article 35 takes up the wish expressed by the UK in the Political Declaration of December 2019, i.e., recognizing the possibility for the UK to design its own emissions trading system, which should have at least "the same scope and effectiveness as that provided" by the EU ETS, and anticipating the possibility that the two systems can be linked.[85]

This likelihood of linking the two emissions trading systems seems to be, at least for the time being, the UK government's first option,[86] even though it is worth noting that the current UK government's approach excludes the principle of nonregression, which is replaced by a series of vague mentions of sustainability in several chapters of its negotiation text. Linking the two emissions trading systems, as expressly referred to in article 25 of Directive 2003/87/EC,[87] would allow participants to use the GHG allowances from one of the systems in the other, so that they can prove compliance with the assigned reduction targets. This will require an international agreement between the two parties, along the lines of the one concluded by the EU with Switzerland in November 2017 (in force from 1 January 2020),[88] which could serve as a reference for these negotiations. In fact, issues related to environmental protection and the fight against climate change, among others, are excluded from what for the UK should be a negotiation of a Comprehensive Free Trade Agreement and are postponed to future supplementary agreements.

6. Final remarks

Over the past few years, the EU has regained some of its leadership in the global fight against climate change, which it had lost in the first decade of the new century due to the fiasco of the Copenhagen Conference. Although the EU is still playing a diminished role in many of the major international debates, it has gradually strengthened its legitimacy in the fight against climate change, not only in the eyes of European citizens, but in the eyes of third countries as well.

But this does not mean that it has fulfilled its ambition to become a major international player on the world stage, and there are still major internal and external constraints on its ability to act as such. Among the former, the most important is probably how to overcome the institutional constraints imposed by the EU system in the negotiation of major international agreements and, in general, in the use of its foreign policy instruments. The sharing of competences between the EU and its member states, or even the distribution of these competences among the EU institutions, with their complex decision-making procedures and major difficulties in defining its external action priorities, does not favour the adoption or promotion of measures of a cross-cutting nature, such as those required to combat climate change. The use of mixed arrangements is necessary to ensure a balance in the exercise of the competences that the EU shares with the member states, but its excessive complexity significantly limits the EU's capacity for action and the scope of its material intervention, while making it difficult to ensure the coherence of its action *ad extra*. In this respect, the role of the EEAS in implementing a more coherent approach to address the various cross-cutting challenges in the fight against climate change should be strengthened.

This challenge is particularly relevant when it comes to measures that aim to integrate the environmental dimension in EU policies and, specifically, in climate change policy. In general terms, it is worth noting that the application of the integration principle has been insufficient and has lacked concreteness at the level of the EU policies concerned, while the EU has failed to internalize the environmental cost. Basically, this principle has only been of practical significance when the EU policy in question and its specific objectives have allowed this. In other words, the principle of integration remains of little real relevance to EU policy-making and is determined by the political will of the actors involved.[89]

It would probably be necessary to state explicitly that non-environmental policies, even at international level, should also pursue environmental objectives, be based on environmental principles, and take account of environmental criteria. This would be important insofar as the various measures that can serve the fight against climate change can be applied within the framework of other sectorial policies. In this regard, the impact assessment reports that the European Commission prepares on legislative initiatives that may have environmental repercussions, or on recommendations for the negotiation of international agreements, although representing a step in the right direction, have proved to be insufficient because they have been considered more as a routine procedural requirement in the preparation of the European Commission's proposal than as a principle of action requiring the application of environmental criteria.[90]

The EU also faces significant exogenous constraints in reaffirming its international leadership in the fight against climate change. On the one hand, the strengthening of bilateral cooperation is particularly important, especially with China. In the medium term, this bilateral cooperation should yield specific actions. The next EU-China summit at the end of 2020 would be a good opportunity to open up a scenario in which the two blocs can reach a bilateral commitment on climate change. However, while year 2020 was to be, at least initially, a turning point for China-EU relations, it remains to be seen whether the consequences of the COVID-19 crisis will lead to a recession or a setback, or if, on the contrary, the EU will try to redress the situation in the hypothetical case that the summit can be held as planned.

The CoP 26 in Glasgow in 2021 will surely be the acid test of this rapprochement between the EU and China, as it will require both sides to take up a clear position regarding the ambition of their new NDCs, which will probably determine the success or failure of the summit. The EU fully supports the increase of the target from -40% to -50%, or even -55%, of GHGs by 2030,[91] without it being entirely clear at this point how the member states will contribute to the achievement of this ambitious goal.[92] On the other hand, the actual contribution of China's new NDCs will be particularly significant,[93] given that they account for more than a quarter of global emissions and that their foreign investments in the fossil fuel sector will also be decisive in terms of the carbon intensity of emissions in many countries.

As far as Brexit is concerned, the EU must provide a clear regulatory framework for its member states and, in particular, for economic operators in the EU by the end of 2020. For now, however, its plans are shrouded in uncertainty. The situation may worsen, both in the economic sectors covered by the EU ETS and in those left out, if the current negotiations do not come to a successful conclusion in the expected time. Although article 132 of the Withdrawal Agreement provides for the possibility that the Joint Committee may extend the transitional period by two years (in which case it must do so before 1 July 2020), clause 33 of the UK's European Union (Withdrawal Agreement) Act 2020 clearly states that "A Minister of the Crown may not agree in the Joint Committee to an extension of the implementation period" and the British government has recently insisted on this point.[94] Therefore, a new period of unrest may open up again in the hypothetical event that the agreement on the future relationship is not reached by the end of the transitional period.[95]

The final exit of the UK in December 2020 without an agreement setting out its future relationship would create great legal uncertainty over the GHG emission allowances generated in its territory. In contrast, the possibility of an orderly agreement, providing for the continued participation of the UK in the EU ETS, as is currently the case of Norway and Iceland, or through its own domestic legislation (as proposed, for example, for financial services) would allow some important guidelines to be maintained in order to ensure joint fulfilment of the mitigation obligations of member states and the UK under the Paris Agreement. However, it seems that for the time being, the UK favours a formula which, without implying permanence in the EU ETS, would allow the linking of its domestic GHG emissions trading scheme with that of the EU from 2020. If that is the case, it is important to ensure, from a technical point of view, that the mutual recognition of the UK's allowances does not undermine the environmental integrity and effectiveness of the system under Directive 2003/87/EC.[96]

A new stage is therefore now opening up in which it will be necessary to articulate these changes in the future relationship between the UK and the EU through new international agreements. The intensity of the consequences of Brexit for the EU will depend on the extent that the UK decides to remain aligned with the EU in this area. For now, the biggest concern remains the process of negotiating the new agreement that will govern future EU-UK relations and whether it will be possible to adopt and ratify this agreement by 31 December 2020. In short, the specific implications for climate change of the UK's exit from the EU are still uncertain.

Notes

1 Oberthür and Pallemaerts, 2010; Giles Carnero, 2017; Bäckstrand and Elgstrom, 2013; Gupta and Grubb, 2000; Parker and Karlsson, 2010; Wurzel and Connelly, 2011; Kulosevi, 2012; Yamin, 2000; Gupta and Ringus, 2001.
2 Council Directive 2003/87/EC of 13 October 2003 on greenhouse gas emission allowance trading system, OJ No. L 275, 25.10.2003 and Council Directive 2018/410/EU to enhance cost-effective emission reductions and low-carbon investments, OJ No. L 76, 19.3.2018.
3 UNTS, vol. 2303, p. 162.
4 Underdal, 1994, p. 183, cited in Schreurs, 2016, p. 220.
5 Parker and Karlsson, 2017, p. 239.
6 UNTS, annex A, No. 54113 (UNTS volume No. has not yet been determined for this record). The Agreement is published at OJ No. L 282, 19.10.2016.
7 Giles Carnero, 2017, p. 213; Falkner, 2016, p. 1111.
8 Draft Council Conclusions on Climate Diplomacy, Brussels 20 January 2020, RELEX.1.C. Through this action, the EU makes the fight against climate change one of its priorities in the field of external relations and reiterates its willingness to lead by example in raising the level of global climate ambition.
9 See Chapter 4.
10 UNTS, vol. 1771, p. 107.
11 Council Decision 2010/427/EU of 26 July 2010 establishing the organization and functioning of the European External Action Service, OJ No. L 201, 3.8.2010.
12 Fajardo del Castillo, 2019, pp. 113–114.
13 Jordan et al., 2010, pp. 39–40, 192, 200–201.
14 Slingenberg, 2004, pp. 21–216; Cremona, 2012, p. 45.
15 Cremona, 2011; Neframi, 2007, p. 15.
16 De Sadeleer, 2012; Jans and Vedder, 2008, pp. 42–43.
17 Marin Duran and Morgera, 2006, p. 11.
18 Opinion 2/00, Cartagena Protocol on the prevention of biotechnological risks, 6 December 2001, ECLI:EU:C:2001:664.
19 Case C-366/10, The Air Transport Association of America et alt. v The Secretary of State for Energy and Climate Change, 21 December 2011, ECLI:EU:C:2011:864.
20 Damro, 2012, p. 70.
21 Fajardo del Castillo, 2010, p. 378.
22 Peeters, 2014, p. 46. While article 192 of the TFEU refers generally to the ordinary legislative co-decision procedure between the Council and the European Parliament, paragraph 2 of that article requires the Council to act unanimously on measures concerning fiscal matters, planning, water management, land use other than waste management, and the choice of energy supplies.

23 Sánchez Galera, 2018, p. 169; Krämer, 2006, pp. 291, 293; Peeters, 2014, pp. 48, 63.
24 Krämer, 2004, p. 33.
25 European Commission Communication: Europe 2020: A strategy for smart, sustainable and inclusive growth, COM(2010)2020 final, 3.3.2010.
26 European Commission Communication: Closing the loop – An EU action plan for the Circular Economy, COM(2015)614 final, 2.12.2015.
27 European Commission Communication: A Roadmap for moving to a competitive low carbon economy in 2050 COM(2011)112 final, 8.3.2011.
28 An essential element in achieving this objective will be the European Commission's new proposal for a Regulation (EU) on the establishment of a framework to facilitate sustainable investment, COM(2018)353 final, 24.5.2018 on which the Council and the European Parliament have already reached political agreement on 7 February 2020 (2018/0178 COD).
29 European Commission Communication: Next steps for a sustainable European future, COM(2016)739 final, 22.11.2016.
30 European Commission Communication: The European Green Deal, COM(2019)640 final, 11.12.2109.
31 European Commission Communication: Commission Work Programme 2019. Delivering what we promised and preparing the future, COM(2018)800 final, 23.10.2018.
32 Kulovesi et al., 2011, pp. 834–836.
33 De Sadeleer, 2010, pp. 526, 531.
34 Cremona, 2012, p. 37; Dhondt, 2003, pp. 80–98.
35 Dhondt, 2003, p. 101.
36 European Commission Communication: Impact Assessment, COM(2002)276 final, 5.6.2002.
37 Diehl et al., 2016, pp. 7–8.
38 Stern, 1986, p. 32; Kulosevi, 2012, p. 140. See also Chapter 4.
39 OJ No. L 275, 25.10.2003.
40 OJ No. L 8, 13.1.2009.
41 Case C-366/10, The Air Transport Association of America et alt. v The Secretary of State for Energy and Climate Change, 21.12.2011. ECLI:EU:C:2011:864.
42 OJ No. L 113, 25.4.2013.
43 OJ No. L 129, 30.4.2014.
44 OJ No. L 350, 29.12.2017.
45 See Chapter 4.
46 Averchenkova et al., 2016, p. 18.
47 UNTC, C.N.575.2019.TREATIES-XXVII.7.d of 4 November 2019 and US Department of State: "On the U.S. Withdrawal from the Paris Agreement. Press Statement", 4 November 2019, available from: www.state.gov/on-the-u-s-withdrawal-from-the-paris-agreement/ [accessed 9 February 2020].
48 Raustiala, 2002, pp. 5, 23, 76; Ahdieh, 2009, pp. 111, 143.
49 International Energy Agency, 2019.
50 Squintati and Vedder, 2014, p. 242; De Matteis, 2010, p. 449.
51 China-EU Partnership on Climate Change Rolling Work Plan, 16 October 2006, available from: www.mfa.gov.cnleng/wjb/zzjg/tyfis/tfsxw/t283051.htm [accessed 9 February 2020].
52 Memorandum of Understanding on Cooperation on Near-Zero Emission Power Generation Technology through Carbon Dioxide Capture and Storage (Phase I), 20 February 2006, available from: ec.europa.eu › files › nzec › docs › nzec_mou_en [accessed 9 February 2020] and Memorandum of Understanding on Cooperation on Near-Zero Emission Coal (NZEC) Power Generation Technology through Carbon Dioxide Capture and Storage (CCS) (Phase II), 30 November 2009, available from: www.fmprc.gov.cn/mfa_eng/wjdt_665385/2649_665393/t630507.shtml [accessed 9 February 2020].

53 Duan, 2015, p. 233; Belis et al., 2015, pp. 212–213; Men, 2014, pp. 54–56.
54 OJ No. L 140, 5.6.2009.
55 Available from: https://www.consilium.europa.eu/es/meetings/internationalsummit/ 2015/06/29/ [accessed 9 February 2020].
56 In 2015, China announced that by 2030 it will achieve the peaking of carbon dioxide emissions, and that it will lower carbon dioxide emissions per unit of GDP by 60% to 65% from the 2005 level. UNFCCC, INDCs as communicated by Parties, available from: https://www4.unfccc.int/sites/submissions/indc/Submission%20Pages/submissions.aspx [accessed 9 February 2020].
57 Belis et al., 2015, p. 209; Pérez de las Heras, 2013, p. 40; Men, 2014, p. 54.
58 Available from: https://ec.europa.eu/info/news/eu-and-china-step-cooperation-climate-change-and-clean-energy-2018-jul-16_en [accessed 9 February 2020].
59 Joint statement of the 21st EU-China summit, available from: www.consilium.europa.eu/en/press/press-releases/2019/04/09/joint-statement-of-the-21st-eu-china-summit/ [accessed 9 February 2020].
60 It is interesting to note that at the same time that the EU is desperately seeking this rapprochement, it considers China, simultaneously "a cooperation partner with whom the EU has closely aligned objectives, a negotiating partner with whom the EU needs to find a balance of interests, an economic competitor in the pursuit of technological leadership, and a systemic rival promoting alternative models of governance", European Commission Comunication: EU-China. A strategic look, JOIN(2019)5 final, 12.3.2019, p. 1. See also, the interview to the High Representative of the Union for Foreign Affairs and Security Policy, of the 2 May 2020 at *Le Journal du Dimanche*, available from: www.lejdd.fr/International/josep-borrel-le-chef-de-la-diplomatie-europeenne-avec-la-chine-nous-avons-ete-un-peu-naifs-3965872 [accessed 10 May 2020].
61 Falkner, 2016, p. 1112.
62 Yu, 2018, p. 252.
63 European Commission: Report of the 27th round of negotiations on the EU-China Comprehensive Agreement on Investment, TRADE.B2/Ares(2020)1434741, 9.3.2020. See also JOIN(2019)5 final, p. 3 in which cooperating with China to fight climate change is one of the ten actions proposed by the Commission.
64 OJ No. L 29, 31.1.2020.
65 Available from: www.legislation.gov.uk/ukpga/2020/1/contents/enacted [accessed 9 February 2020].
66 The only previous relevant case of withdrawal is that of Greenland. However, the borders of Denmark were not changed. Rather, by unilateral decision of Denmark as a EU member state, the application of EU law was excluded from territories that continued under its sovereignty. Founding Treaties were modified in 1984, formally excluding Greenland from their territorial scope and including this territory as an overseas territory in Annex IV of the European Economic Community Treaty since February 1985 (Treaty amending, with regard to Greenland, the Treaties establishing the European Communities Protocol on special arrangements for Greenland, OJ No. L 29, 1.2.1985).
67 The Climate Change Act 2008 (2050 Target Amendment) Order 2019 (S.I. 2019/1056), available from: www.legislation.gov.uk/ukpga/2008/27 [accessed 9 February 2020].
68 Council Regulation (EU) 2018/842 of 30 May 2018 on binding annual greenhouse gas emission reductions by Member States from 2021 to 2030 contributing to climate action to meet commitments under the Paris Agreement, OJ No. L 156, 19.6.2018. *Vid.*, *also* EU Commission Communication: A policy framework for climate and energy in the period from 2020 to 2030, COM(2014)15 final, 22.1.2014; Conclusions of the European Council of 23/24 October 2014, EUCO 169/14.
69 Council Regulation (EU) 525/2013 of 21 May. 2013 on a mechanism for monitoring and reporting greenhouse gas emissions and for reporting other information at national and Union level relevant to climate change, OJ No. L 165, 18.6.2013.

70 Council Regulation (EU) 2019/631 of 17 April 2019 setting CO_2 emission performance standards for new passenger cars and for new light commercial vehicles, OJ No. L 111, 25.4.2019.

71 Council Decision No 406/2009/EC of 23 April 2009 on the effort of Member States to reduce their greenhouse gas emissions to meet the Community's greenhouse gas emission reduction commitments up to 2020, OJ No. L 140, 5.6.2009.

72 Council Regulation (EU) 2017/2392 of 13 December 2017 amending Directive 2003/87/EC to continue current limitations of scope for aviation activities and to prepare to implement a global market-based measure from 2021, OJ No. L 350, 29.12.2017.

73 European Commission, Directorate-General Climate Action: Notice to Stakeholders Withdrawal of the United Kingdom and the EU Emissions Trading System (ETS), Brussels, 19 December 2018.

74 Commission Regulation (EU) 2018/208 of 12 February 2018 amending *Regulation (EU)* No 389/2013 establishing a Union Registry, OJ No. L 39, 13.2.2014.

75 Dupont and Moore, 2019, p. 55; Bocse, 2020.

76 Macrory and Newbigin, 2018, p. 246. In fact, paragraph 4 of article 2 of Annex 4 of the Withdrawal Agreement clearly stresses this continuity in the international commitment and indicates that "The Union and the United Kingdom shall take the necessary measures to meet their respective commitments to international agreements to address climate change, including those which implement the United Nations Framework Conventions on Climate Change, such as the Paris Agreement of 2015".

77 The Climate Change Act 2008 (2050 Target Amendment) Order 2019 (S.I. 2019/1056), available from: www.legislation.gov.uk/ukpga/2008/27 [accessed 9 February 2020].

78 Saint-Genies, 2018, p. 157; Marcantoui, 2016.

79 Averchenkova et al., 2016, p. 9; Saint-Genies, 2018, p. 159.

80 European Environmental Agency, 2019, available from: www.eea.europa.eu/data-and-maps/indicators/greenhouse-gas-emission-trends-6/assessment-3 [accessed 9 February 2020]. See also, Committee on Climate Change, Reducing UK emissions, 2019 Progress Report to Parliament, 10 July 2019, available from: www.theccc.org.uk/publication/reducing-uk-emissions-2019-progress-report-to-parliament/ [accessed 9 February 2020].

81 e.g., Commonwealth Secretariat, Commonwealth Secretariat Strategic Plan 2017/18–2020/21 (1 June 2017), available from: https://thecommonwealth.org/our-work/climate-change [accessed 9 February 2020].

82 See the composition of those Groups at https://unfccc.int/process-and-meetings/parties-non-party-stakeholders/parties/party-groupings [accessed 9 February 2020].

83 Political declaration setting out the framework for the future relationship between the European Union and the United Kingdom, OJ No. C I384, 11.12.2019.

84 European Commission, Task Force for Relations with the United Kingdom, Draft text of the Agreement on the New Partnership with the United Kingdom, 18 March 2020 UKTF /2020)14.

85 Hinson and Priestley, 2019, p. 38.

86 HM Government, The Future Relationship with the EU the UK's Approach to Negotiations, Presented to Parliament by the Prime Minister, CP211 (February 2020), pp. 22–23.

87 Article 25 allows those agreements to be concluded "with third countries listed in Annex B to the Kyoto Protocol which have ratified the Protocol to provide for the mutual recognition of allowances between the EU ETS and other greenhouse gas emissions trading systems". Those may provide for "the recognition of allowances between the EU ETS and compatible mandatory greenhouse gas emissions trading systems with absolute emissions caps established in any other country or in sub-federal or regional entities".

88 Decision No 2/2019 of the Joint Committee established by the Agreement between the European Union and the Swiss Confederation on the linking of their greenhouse gas Emissions trading systems of 5 December 2019 amending annexes I and II to the Agreement between the European Union and the Swiss Confederation on the linking of their greenhouse gas emissions trading systems; and Council Decision (EU) 2019/2106 of 21 November 2019 on the position to be taken, on behalf of the European Union, within the Joint Committee established by the Agreement between the European Union and the Swiss Confederation on the linking of their greenhouse gas emissions trading systems, as regards the amendment of Annexes I and II to the Agreement, OJ No. L 318, 10.12.2019.

89 Dhondt, 2003, p. 483.

90 Diehl et al., 2016, pp. 7–8, 14.

91 COM(2019)640 final.

92 UNFCCC, INDCs as communicated by Parties, available from: https://www4.unfccc. int/sites/submissions/indc/Submission%20Pages/submissions.aspx [accessed 9 February 2020] and COM(2014)15 final; European Commission Communication: A European strategic long-term vision for a prosperous, modern, competitive and climate neutral economy, COM(2018)773 final, 28.11.2018; Conclusions of the European Council of 23/24 October, EUCO 169/14; Resolution of the European Parliament on Climate and Environmental Emergency, 28.11. 2019 (provisional edition), P9_TA-PROV(2019)0078.

93 Vallejo, 2019.

94 HM Government, The Future Relationship with the EU. The UK's Approach to Negotiations, Presented to Parliament by the Prime Minister, CP211 (February 2020), p. 4.

95 Simson and Garner, 2020.

96 Campins, 2010, p. 33.

References

Ahdieh, R.B., 2009. Foreign Affairs, International Law and the New Federalism: Lessons from Coordination. *Constitutional Theory and Reality*, 24, 109–172.

Averchenkova, A., Bassi, S., Benes, K.J., Green, F., Lagarde, A., Neuweg, I. and Zachmann, G., 2016. *Climate Policy in China, the European Union and the United States: Main Drivers and Prospects for the Future*. The Grantham Research Institute on Climate Change and the Environment. Available from: www.lse.ac.uk/granthaminstitute.

Bäckstrand, K. and Elgström, O., 2013. The EU's Role in Climate Change Negotiations: From Leader to "Leadiator". *Journal of European Public Policy*, 20(10), 1369–1386.

Belis, D., Joffe, P., Kerremans, B. and Qi, Y., 2015. China, the United States and the European Union: Multiple Bilateralism and Prospects for a New Climate Change Diplomacy. *Carbon & Climate Law Review*, 3, 203–218.

Bocse, A.M., 2020. The UK's Decision to Leave the European Union (Brexit) and Its Impact on the EU as a Climate Change Actor. *Journal Climate Policy*, 20(2), 265–274.

Campins Eritja, M., 2010. Las entidades sub-nacionales en Norteamérica y la lucha contra el cambio climático: desarrollo normativo y vinculación de sistemas de comercio de derechos de emisión. *Revista Catalana de Dret Ambiental*, 1(2), 1–48. Available from: https://revistes.urv.cat/index.php/rcda/issue/view/87.

Cremona, M., 2011. External Relations and External Competence: The Emergence of an Integrated Policy. In Craig, P. and De Burca, G. (Editors), *The Evolution of EU Law*. Oxford: Oxford University Press, 217–268.

Cremona, M., 2012. Coherence and EU External Environmental Policy. In Morgera, E. (Editor), *The External Environmental Policy of the European Union*. Cambridge: Cambridge University Press, 33–54.

Damro, C., 2012. The Post-Lisbon Institutions and EU External Environmental Policy. In Morgera, E. (Editor), *The External Environmental Policy of the European Union*. Cambridge: Cambridge University Press, 55–75.

De Matteis, P., 2010. EU – China Cooperation in the Field of Energy, Environment and Climate Change. *Journal of Contemporary European Research*, 6, 449–477.

De Sadeleer, N., 2010. *Environment and Internal Market. Commentary J. Megret*, 3rd ed. Brussels: Editions de l'Université de Bruxelles.

De Sadeleer, N., 2012. Principle of Subsidiarity and the EU Environmental Policy. *Journal for European Environmental & Planning Law*, 9(1), 63–70.

Dhondt, N., 2003. *Integration of Environmental Protection into other EC Policies. Legal Theory and Practice*, The Avosseta Series 2. Groningen: Europa Law Publishing.

Diehl, K., Burkhard, B. and Jacob, K., 2016. Should the Ecosystem Services Concept Be Used in European Commission Impact Assessment? *Ecological Indicators*, 61, 6–17.

Duan, M., 2015. From Carbon Emissions Trading Pilots to National System: The Road Map for China? *Carbon & Climate Law Review*, 9(3), 231–242.

Dupont, C. and Moore, B., 2019. Brexit and the EU in Global Climate Governance. *Politics and Governance*, 7(3), 51–61.

European Environmental Agency, 2019. *Total Greenhouse Gas Emission Trends and Projections in Europe*, 19 December 2019. Available from: www.eea.europa.eu/data-and-maps/indicators/greenhouse-gas-emission-trends-6/assessment-3.

Fajardo del Castillo, T., 2010. Revisiting the External Dimension of the Environmental Policy of the European Union: Some Challenges Ahead. *Journal for European Environmental and Planning Law*, 7(4), 365–390.

Fajardo del Castillo, T., 2019. Competencia Exterior Medioambiental de la Unión Europea y Desarrollo Progresivo del Derecho Internacional en el Marco de la Asamblea General de Naciones Unidas. *Revista General de Derecho Europeo*, 47, 110–158.

Falkner, R., 2016. The Paris Agreement and the New Logic of International Climate Politics. *International Affairs*, 92(5), 1107–1125.

Giles Carnero, R., 2017. The European Union's Contribution to the Development of the International Climate Change Regime: The European Climate and Energy Package in the Context of International Action. *Cuadernos Europeos de Deusto*, 57, 193–215.

Gupta, J. and Grubb, M.J. (Editors), 2000. *Climate Change and European Leadership. A Sustainable Role for Europe?* Dordrecht: Kluwer Academic Publishers.

Gupta, J. and Ringius, L., 2001. The EU's Climate Leadership: Reconciling Ambition and Reality. *International Environmental Agreements: Politics, Law and Economics*, 1(2), 281–299.

Hinson, S. and Priestley, S., 2019. *Brexit: Energy and Climate Change*, Briefing Paper Number CBP 8394. London: House of Commons.

International Energy Agency, 2019. *World Energy Outlook 2019*. Available from: www.iea.org/reports/world-energy-outlook-2019.

Jans, J. and Vedder, H., 2008. *European Environmental Law*. Dordrecht: Europa Law Publishing.

Jordan, A., Huitema, D., van Asselt, H., Rayner, T. and Berkhout, F. (Editors), 2010. *Climate Change Policy in the European Union: Confronting the Dilemmas of Mitigation and Adaptation?* Cambridge: Cambridge University Press.

Krämer, L., 2004. The Genesis of EC Environmental Principles. In Macrory, R. (Editor), *Principles of European Environmental Law*. Dordrecht: Europa Law Publishing, 33–35.

Krämer, L., 2006. Some Reflections on the EU Mix of Instruments on Climate Change. In Peeters, M. and Deketelaere, K. (Editors), *EU Climate Change Policy. The Challenge of New Regulatory Initiatives*. Cheltenham: Edward Elgar Publishing, 279–296.

Kulosevi, K., 2012. Climate Change in EU External Relations: Please Follow My Example (or I Might Force You to). In Morgera, E. (Editor), *The External Environmental Policy of the European Union*. Cambridge: Cambridge University Press, 115–148.

Kulovesi, K., Morgera, E. and Muñoz, M., 2011. Environmental Integration and Multi-faceted International Dimensions of EU Law: Unpacking the EU's 2009 Climate and Energy Package. *Common Market Law Review*, 48(3), 829–891.

Macrory, R. and Newbigin, J., 2018. Brexit and International Environmental Law. In Fitzgerald, O.E. and Lein, E. (Editors), *Complexity's Embrace. The International Law Implications of Brexit*. Montreal: McGill-Queen's University Press, 241–252.

Marcantouiui, C., 2016. *The Impact of Brexit on Climate Policy: The EU and the Paris Agreement Florence School of Regulation*. Florence: European University Institute. Available from: https://fsr.eui.eu/eu-paris-agreement/.

Marin Duran, G. and Morgera, E., 2006. The UN 2005 World Summit, the Environment and the EU: Priorities, Promises and Prospects. *Review of European Community and International Environmental Law*, 15(1), 11–22.

Men, J., 2014. Climate Change and EU – China Partnership: Realist Disguise or Institutionalist Blessing? *Asia Europe Journal*, 12, 49–62.

Neframi, E., 2007. *Research on Mixed Agreements of the European Community. Community and International Aspects*. Brussels: Bruylant.

Oberthür, S. and Pallemaerts, M. (Editors), 2010. *The New Climate Policies of the European Union: Internal Legislation and Climate Policy*. Brussels: Institute for European Studies, VUBPRESS.

Parker, C.F. and Karlsson, C., 2010. Climate Change and the European Union's Leadership Moment: An Inconvenient Truth? *Journal of Common Market Studies*, 48(4), 923–943.

Parker, C.F. and Karlsson, C., 2017. Assessing the European Union's Global Climate Change Leadership: From Copenhagen to the Paris Agreement. *Journal of European Integration*, 239–252.

Peeters, M., 2014. Governing Towards Renewable Energy in the EU: Competences, Instruments, and Procedures. *Maastricht Journal of European and Comparative Law*, 21(1), 39–63.

Perez de las Heras, B., 2013. The European Union, the United States, and China Dialogue on Climate Change: Respective Policies and Mutual Synergies for a World Climate Order. *Georgetown International Environmental Law Review*, 26, 13–46.

Raustiala, K., 2002. The Architecture of International Cooperation: Transgovernmental Networks and the Future of International Law. *Virginia Journal of International Law*, 43, 1–92.

Saint-Genies, G., 2018. À propos de certaines implications juridiques du BREXIT dans le domaine de la lutte contre les changements climatiques. *Revue Quebecoise de Droit International* (Special Issue), 147–164.

Sánchez Galera, M.D., 2018. The EU Sustainability Model at the Crossroads of a Leadership on Climate Policy and the Failure of a Strong Energy Integration. A Fragmented European Scenario. In Giles Carnero, R. (Editor), *Desafíos de la acción jurídica internacional y europea frente al cambio climático*. Barcelona: Atelier, 155–171.

Simson Caird, J. and Garner, O., 2020. *The European Union (Withdrawal Agreement) Bill and the Rule of Law*. Available from: https://binghamcentre.biicl.org/publications/the-european-union-withdrawal-agreement-bill-and-the-rule-of-law.

Squintatl, L. and Vedder, H.H.B. (Editors), 2014. *Sustainable Energy. United in Diversity. Challenges and Approaches in Energy Transition in the European Union*. Leipzig: European Environmental Law Forum Book Series.

Slingenberg, Y., 2004. Community Action in the Fight Against Climate Change. In Onida, M. (Editor), *Europe and the Environment. Legal Essays in Honour of L. Kramer*. Groningen: Europa Law Publishing, 209–227.

Stern, B., 1986. Quelques observations sur les règles internationales relatives à l'application extraterritoriale du droit. *Annuaire Français de Droit International*, 32(1), 7–52.

Underdal, A., 1994. Leadership Theory: Rediscovering the Arts of Management. In Zartman, I.W (Editor), *International Multilateral Negotiation: Approaches to the Management of Complexity*. San Francisco, CA: Jossey-Bass Publishers, 178–197. Cited in Schreurs, M.A., 2016. The Paris Climate Agreement and the Three Largest Emitters: China, the United States, and the European Union. *Politics and Governance*, 4(3), 219–223.

Vallejo, L., 2019. *COP 25 Must Shape the Strengthening of Collective Ambition, 2019*. Paris: IDDRI. Available from: www.iddri.org/fr/publications-et-evenements/billet-de-blog/la-cop-25-doit-faconner-le-renforcement-de-lambition.

Wurzel, R.K.W. and Connelly, J. (Editors), 2011. *The European Union as a Leader in International Climate Change Politics*. London: Routledge.

Yamin, F., 2000. The Role of the EU in Climate Negotiations. In Gupta, J. and Grubb, M. (Editors), *Climate Change and European Leadership: A Sustainable Role for Europe?* Dordrecht: Kluwer, 47–66.

Yu, K., 2018. Energy Cooperation in the Belt and Road Initiative: EU Experience of the Trans-European Networks for Energy. *Asia Europe Journal*, 16, 251–265.

3   The EU diplomacy for biodiversity

Teresa Fajardo del Castillo

1. Introductory remarks

The year 2020 was destined to be the "super year for biodiversity".[1] However, the 15th Conference of the Parties of the Convention on Biological Diversity[2] (CBD CoP 15, hereinafter) that was to define the way forward for global cooperation over the next decade had to be postponed due to the coronavirus pandemic. The coronavirus, which has spread all over the planet, had its origin in a problem related to biodiversity. A case of zoonosis, the transmission of disease from animals to humans was triggered by human interference with wildlife and is now transforming the world as we know it. Thus, the next EU Biodiversity Strategy to 2030[3] will face both threats and opportunities, and the EU, now more than ever, must take on the role of the champion of green multilateralism[4] by supporting the adoption of a post-2020 global biodiversity framework.[5]

The EU has already aligned with the United Nations (UN) proposals presented in the Zero Draft of the CBD post-2020 global biodiversity framework.[6] When the present chapter was completed, the pre-conferences leading to the CBD CoP 15 had already been postponed due to the coronavirus crisis; however, several other key events related to the legal mosaic of biodiversity have already been held. At these meetings, the EU has advocated placing biodiversity "high on the political agenda" and stressed that the

> Post-2020 global biodiversity framework must set out an ambitious plan to implement broad-based action to bring about a transformation in society's relationship with biodiversity and to ensure that, by 2050, the shared vision of living in harmony with nature is fulfilled.[7]

The EU and its member states have also manifested their commitment to lead the Summit on biodiversity at the level of heads of state and government that will take place ahead of the CBD CoP 15 in September 2020, as one of their priorities at the United Nations.[8]

The Intergovernmental Science-Policy Platform on Biodiversity and Ecosystem Services Global Assessment (IPBES 2019) reported that around one million species face extinction. This IPBES Global Assessment confronted the EU and its member states with the origin of the problems: "worldwide erosion of biodiversity, caused primarily by changes in how land and sea are used, direct exploitation of natural resources, and with climate change as the third most important driver of biodiversity loss".[9] This diagnosis requires a responsible answer from green multilateralism. Thus, this new decade presents many challenges for the EU as a global actor that has always tried to lead by virtuous example at the UN institutions. From the very start, the EU and its member states supported the United Nations Environment Programme (UNEP) because they understood that the UN is "the habitat for addressing environmental issues that are too large to be handled by any state, or even by a limited group of states".[10] Today, the EU must go on funding UN institutions and initiatives and must take the lead to inspire a better understanding of the problems as well as the solutions, at the risk of being accused of delocalizing its demands of consumption or imposing its rules beyond its borders. Through the external dimension of its environmental policy, the EU and its member states have made an enormous effort to lead global negotiations (which met with success in the case of the Paris Agreement on Climate Change), and they will seek to repeat the same formula in other sectors, such as water and biodiversity. Thus, in 2020, the EU and its member states will try to summon the consensus required to further develop the Convention on Biological Diversity (CBD) and its protocols, supporting the adoption of the new post-2020 global biodiversity framework at the CBD CoP 15. Before that meeting, the EU and its member states must adopt their own strategy, the EU 2030 Biodiversity Strategy, to carry on the work of its earlier Biodiversity Strategy to 2020, adapted to comply with the CBD.[11] However, the targets set out in this strategy are a long way from being met and need to be reformulated without delay, incorporating stronger commitments in view of the poor performance achieved so far.[12]

As a global actor, the EU has been witness to the profound effect and the transformative capacity of multilateral environmental agreements in its domestic policies and the legal instruments that were adopted to apply them at internal level.[13] The EU has also carried out an external action to address global environmental issues to complement its internal dimension and to promote its own standards of protection. Thus, it leads by virtuous example in the field of the protection of biodiversity in Europe through the results of Natura 2000 Network, which has become a pan-European network extended through the Neighbourhood Policy and the pre-accession instruments. The greatest achievements of the EU external action to protect the environment are the successive enlargements that "have extended high standards of environmental protection across a large part of the European continent, and the Union's efforts have contributed to increasing international commitment to combatting climate change and biodiversity loss".[14]

The implementation of the Habitats and Birds Directives[15] has shown that spatial planning can serve to achieve the goal of creating areas of protection, as foreseen in the Aichi Targets[16] and the Sustainable Development Goals.[17] Moreover, the fact that these goals are quantifiable facilitates surveillance of the compliance of the action to fight biodiversity loss both at land and at sea. Nevertheless, the EU also faces problems in the lack of implementation of environmental legislation[18] that directly affects biodiversity. Therefore, the EU and its member states are not addressing some of the drivers of biodiversity loss. The Green Deal acknowledges this failure when it states that "more than 90% of biodiversity loss and water stress come from resource extraction and processing of materials, fuels and food".[19] Moreover, the dilution of conservationist policies through the member states is placing the natural capital both of Europe and of the entire world at risk.[20] In recent years, the European Environment Agency has reported that the number of protected terrestrial and marine areas in the EU member states has increased,[21] but it also notes the existence of a "predominantly unfavourable status of conservation at 60% for species and 77% for habitats".[22] This shows that real and effective protection requires much more than the creation of more areas, protected not just on paper but through better management and compliance rates.[23]

The purpose of this chapter is to examine how EU policies and legal tools can bolster the goals of a new EU 2030 Biodiversity Strategy at global level and contribute to the consolidation of a EU diplomacy for biodiversity. A new strategy will also aim to deliver the goal of the European Green Deal adopted by the European Commission in 2019 and endorsed by the Council and the European Parliament,[24] namely "to transform the EU's economy for a sustainable future".[25] This Green Deal will have an external dimension that will test the EU's capacity to lead the fight against biodiversity loss all over the globe. Thus, the second section examines the EU's competence to protect biodiversity globally and the external actions designed to this end. The third section explores whether the EU's external action in the field of biodiversity protection might lead to the emergence of diplomacy for biodiversity as a distinct concept inside the framework of the EU Green Diplomacy. The fourth section will present recommendations for a new EU 2030 Biodiversity Strategy to align with Green Multilateralism, as promoted by the UN institutions and the CBD CoP. The final remarks discuss state-of-the-art global biodiversity strategies in the era of coronavirus.

2. A EU competence and a legal mandate to protect biodiversity globally

The EU's external environmental action has a very special profile because it was implemented in the exercise of one of its external competences expressly foreseen in the treaties together with the implicit competence derived from the exercise of internal competences in accordance with the principle of parallelism of internal and external competences (*in foro interno, in foro externo*).[26] Moreover, one of the main goals of the EU environmental action is to "promot[e] measures at international level to deal with regional or worldwide environmental problems".[27] However, this express competence is also regarded as shared and "mixed": that is, in the external fora and conferences of the parties, the environmental competence shall be exercised by both the EU and member states.[28] This definition of the competence as being shared and mixed has led to numerous institutional conflicts over its nature and extent, the choice of the legal bases, and the representation in international negotiations, conferences, and institutions. Since the Lisbon Treaty, the European Commission has held that it alone should represent the EU in the exercise of this competence. This claim triggered a reaction from the Council and member states which consider that a function such as representation cannot determine the nature of competence or the extent of the referral; they hold that in cases in which the shared competence has not been exercised through the adoption of common rules, a mixed approach should prevail with regard to both the representation and exercise of the external competence. Despite the ongoing institutional disputes on the legal bases justifying the EU's external action and representation, the EU institutions and its member states agree on the common ground of shared interests. From this, a variety of problems arise for the consolidation of EU Green Diplomacy and, as far as this chapter is concerned, of EU diplomacy for biodiversity.

In the pre-Lisbon period, the mixed nature of the competence led to a mixed representation in international institutions and conferences which was at odds with the exclusive function of representation granted to the European Commission by the Lisbon Treaty, especially in those international organizations that granted membership to the EU.[29] In those instances in which the EU has been assigned the lesser status of observer, the formula of "declaration on behalf of the Union" does not reflect its growing competence and external personality. This is still the case in the Antarctica Council, where the protection of biodiversity has been left in the hands of member states. In the context of the institutions of the Antarctic Convention, the European Union Court of Justice (EUCJ) has accepted that the member states remain competent to participate in the bodies in which the EU has not yet been granted a seat.[30]

The institutional conflict between the European Commission and the Council and the EU member states also extends to the exercise of those shared competences that have not yet been widely regulated by the Union, as well as those that the member states believe continue to be part of their sovereign rights, especially in areas in which the progressive development of international law is at issue. In those cases, some member states may request to be involved in the negotiations at the same level with the Union, because invoking the mixed nature of the competence will mean that none of them is competent to conclude the agreement alone. In the case of the CBD, the EU declared its competence to conclude the agreement based on the environmental legislation and policy adopted to protect biodiversity and in the framework of other sectorial policies in Annex B of its decision to conclude the convention,[31] while in Annex C, the EU and its member states insisted on the mixed nature of the competence to negotiate on biotechnology.[32]

The growing concern for biodiversity is also reflected in the Seventh Environment Programme (7EAP)[33] which established a clear mandate – to be considered as a source of law by both European institutions and member states, according to Kramer[34] – "to halt the loss of biodiversity and the degradation of ecosystem services in the Union by 2020, and restore them in so far as feasible, while stepping up the Union contribution to averting global biodiversity loss".[35] This mandate, adopted by both the EU and its member states, still underpins the exercise of the external action in the field of biodiversity, and in most cases still determines the negotiation of mixed agreements.

The progressive development of international law regarding biodiversity, for example the future adoption of a convention on the protection of marine biodiversity beyond areas of national jurisdiction, will require a mixed representation and exercise of the competences as foreseen expressly in the 7EAP, when it states:

> The Union and its member states should proactively engage in international negotiations on new and emerging issues, in particular on new Conventions, agreements and assessments, and, accordingly, reaffirm their strong determination to continue efforts to launch, as soon as possible, negotiations in the framework of a UN General Assembly for an UNCLOS (United Nations Convention on the Law of the Sea) implementing agreement on the conservation and sustainable use of marine biological diversity of Areas Beyond National Jurisdiction and supporting the completion of the first 'World Ocean Assessment.[36]

However, in the rest of the topics, the nature of the agreement will be determined by the exercise of the competence at the internal level and the goals pursued in the EU environmental policies as foreseen in the new legal bases of the Lisbon Treaty.[37] The biodiversity negotiations to be held in the coming years are representative of this conflict between the competences already acquired by the EU and those that are still under a process of progressive development in international institutions. The adoption of an EU 2030 Biodiversity Strategy might exercise a *vis attractiva* for goals to be pursued at the internal as well as the external level, and might become the mandate that the EU needs in order to go further in its support of a global policy to protect biodiversity. However, the mandate under target 6 of the previous Strategy 2020 was not strong enough to "help avert global biodiversity loss" by 2020, despite the EU's efforts to step up its contribution. The Clearing-House Mechanism of the Convention on Biological Diversity that examined the EU achievements acknowledged "progress towards target but at an insufficient rate".[38]

3. Is there a specific EU diplomacy for biodiversity?

The idea of EU Green Diplomacy to protect the environment was first associated with the Green Diplomacy Network (GDN) that was established at the European Council in Thessaloniki in 2003[39] to mandate the rotating presidency to carry out the defence of demarches in the environmental field. The term "demarche" is borrowed from the terminology of classical diplomacy. As Nicolson said "the closest English equivalent for the expression '*faire une démarche*' is 'make representations', but it should be remembered that the word in French covers all manner of representations from proposals to threats".[40] All sorts of demarches appear on the agendas of the EU presented to the UN institutions and multilateral conferences in the different sectors of environmental protection. Thus, EU demarches foresee different goals in the negotiation processes of green multilateralism and, in particular, of biodiversity: from defining goals and agendas to fostering spatial planning and restoration policies, and protecting biodiversity through regional cooperation, as analysed in the following.

Before the Lisbon Treaty, the GDN worked in accordance with a diplomatic method on the basis of demarches that every rotating presidency defended before international organizations and conferences of the parties, as well as in bilateral agreements and relations. Every six months, the GDN gave member states of the rotating presidency the opportunity to lead proposals that matched their domestic interests as well as those shared with the EU. After the Lisbon Treaty, this inter-governmental network and its tasks were integrated into the European External Action Service (EEAS). At the very beginning, the EEAS lacked the diplomatic means it required. Even though it was clear that the diplomatic service of the EU would have an autonomous existence and would have a hybrid staff comprised of civil servants from the European Commission and diplomats from the member states, for the sake of maintaining previous alliances, connections, and know-how, the EEAS had to maintain the diplomatic tradition of the states at the service of the EU interests and goals.

To deliver environmental goals, the EEAS works with the European Commission services because they share crosscutting issues such as cooperation with development, trade, and enlargement policies and the European Neighbourhood Policy. However, they did not develop a diplomatic method that shaped EU negotiations in a recognizable way until the negotiations of the Paris Agreement on climate change. The demarches on climate change laid the foundations for diplomacy on climate change and enhanced the EU's role and visibility. In the case of biodiversity, special diplomacy in this sector is difficult to identify since biodiversity has always played a gregarious role with respect to sustainable development or climate change. Nevertheless, certain features personalize EU diplomacy for biodiversity as informed by goals and principles that come from a multi-layered framework system that the European institutions and their member states have promoted in their external actions. Thus, the so-called environmental conditionality of EU external action makes the granting of European support to developing countries dependent on their acceptance of green multilateralism and multilateral environmental agreements in the field of biodiversity.

The goals have been provided by the 7EAP and the EU Biodiversity Strategy to 2020, as mentioned previously. Target 6 of this strategy was destined to "Help stop the loss of global biodiversity" and provided four specific actions that have been carried out through the diplomacy on biodiversity and external action in the past 10 years.[41] These goals align with the Aichi Targets and the Strategic Plan for Biodiversity 2011–2020, and support the development and enforcement of the CBD and its protocols globally.

These goals are also incorporated in the New Consensus on Development – *Our World, Our Dignity, Our Future* – which is the common framework for development policy adopted by the EU institutions and the member states in which biodiversity appears interlinked with other sectorial policies such as agriculture and climate change. It establishes clearly (though broadly) a general commitment addressing both the goals of conservation and sustainable use of biodiversity and ecosystems, highlighting the contradictory approach from which nature is always observed from the anthropocentric angle of exploitation:

> The EU and its member states will support the conservation and sustainable management and use of natural resources, and the conservation and sustainable use of biodiversity and ecosystems, including forests, oceans, coastal areas, river basins and other ecosystems, for the provision of ecosystem services. In line with international commitments, they will tackle illegal logging and its associated trade, land and forest degradation, desertification, drought, and biodiversity loss.[42]

The principles of diplomacy on biodiversity are those that inform both the EU's internal policies and its external actions: the principle of environmental integration in all relevant policy areas as essential to reducing the pressures on the environment resulting from the policies and activities of other sectors,[43] the adoption of measures based on scientific evidence, and the application of a sustainable impact assessment to all measures that might affect the environment. Approaches such as political strategies and spatial planning, rather than principles with legal status, are applied to mainstreaming essential purposes in sectorial policies and are foreseen in the 7EAP when it states:

> Ecosystem-based approaches to climate change mitigation and adaptation which also benefit biodiversity and the provision of other ecosystem services should be used more extensively as part of the Union's climate change policy, while other environmental objectives such as biodiversity conservation and the protection of soil and water should be fully taken into account in decisions relating to renewable energy.[44]

Environmental principles from the legal spectrum of treaties and the case law of the World Trade Organization (WTO) also inform the trade-related environmental measures that the EU has adopted to protect biodiversity.[45] The EU also fosters avant-garde principles for protecting biodiversity when it proposes transnational cooperation based on "long-lasting engagement . . . in conserving and restoring the ecological connectivity and integrity of ecosystems to support the natural movements of animals, necessary for their survival and well-being".[46] The EU and its member states are also adopting the formulas that were successful in the negotiation of the Paris Agreement to biodiversity agreements in the framework of the discussions on promoting and improving their implementation. Thus, the EU favours the adoption of a bottom-up approach that allows states to establish their national contributions to the protection of biodiversity.[47]

In further research on the existence and extent of consolidation of diplomacy for biodiversity, it will be necessary to examine which of these goals and principles have been accepted in international agreements and policies in this field. If they have not been embraced in international negotiations led by global consensus, then the EU would have failed as a regulatory power and it should reconsider, once more, the diplomatic methods and the external action toolbox it uses to implement what is known as environmental conditionality.[48]

The environmental conditionality of the EU's external action makes cooperation and the granting of aid to development funds dependent on the adherence to the goals and principles enshrined in clauses of cooperation agreements as part of a "softer leadership"[49] that becomes stronger when promoting accession to multilateral environmental agreements – in particular those related to biodiversity – through unilateral trade-related environmental measures.[50] Environmental conditionality as a tool for the protection of biodiversity may be expressed in many ways: from the blandest statements made at global events of green multilateralism to the strongest demands made of candidates for cooperation for development or applicants for EU membership. Clauses pledging to protect the environment and biodiversity are also negotiated and introduced in partnerships and trade agreements with the goal of fostering a common playing field for biodiversity protection that will lead to an environmental rule of law. All are different ways of mainstreaming biodiversity in political dialogue and external action, in particular, in the cooperation for development policy. A review of all these instruments highlights the following shared elements and purposes that implement Target 6 of the EU Biodiversity Strategy to 2020 and its actions, in particular Action 18 destined to increase funding on global biodiversity:

- All the instruments of the external action identify the conservation and sustainable management of natural resources and biodiversity as a common goal of cooperation.
- Development cooperation instruments incorporate funding for "capacity building for participating in and implementing multilateral environmental agreements, including biodiversity, biosafety and CITES". The goal of contributing to capacity building in developing countries appears in bilateral and multilateral agreements,[51] as well as in the resolutions adopted in the CoP of the Cartagena Protocol on biosafety and the Nagoya Protocol on access and benefit sharing.[52]
- At the regional level, as in the case of Africa and Latin America, the EU encourages harmonization of environmental legislation with greater scope and ambition to promote formulas of cooperation and integration.[53]
- In all agreements, the EU promotes control mechanisms to enhance their implementation.
- The EU funds biodiversity projects carried out in partnership with the United Nations Development Programme such as BIOFIN, the Biodiversity Financing Initiative.[54]
- The EU finances biodiversity through thematic instruments such as the "B4Life flagship initiative", the EU Biodiversity for Life Project.

These measures are foreseen in cooperation framework agreements or even in political dialogues that the EU and its member states engage in with third countries and regional groups in order to consolidate a legally articulated system and political understanding on common concerns. In addition, on an unilateral basis, the EU regularly incorporates provisions in the EU's Free Trade Agreements with third countries and uses trade policy instruments to improve the implementation of multilateral environmental agreements such as the Convention on the Legal Trafficking of Endangered Species of Wild Flora and Fauna (CITES).[55] Developing countries benefit from additional trade advantages through the Generalized System of Preferences: the system of trade incentives granted to states willing to adhere to UN conventions in areas such as human and labour rights, environmental protection,[56] good governance, and the fight against drugs, which are well established in practice but still present room for improvement, as shown in recent studies that have assessed the capacity of this scheme to enhance compliance and implementation of international agreements by creating a GSP Compliance Index.[57] The EU Commission's last assessment on the results achieved by this scheme highlighted that "with regard to environment and climate change, countries have improved reporting".[58]

The Achilles heel of environmental conditionality and, in general, of external action to protect biodiversity is the lack of control and monitoring on the results obtained with the different instruments used to pursue Target 6 of the EU Biodiversity Strategy to 2020. The reasons for this deficiency are manifold: a chronic absence of data on the third states performance, insufficient control by the EU delegations on the spot, insufficient human and economic resources, and a weak governance system at national and global level which waste the opportunities to integrate biodiversity protection in the decision-making process at all political levels.[59]

4. Recommendations for the EU 2030 Biodiversity Strategy: aligning its objectives with those of Green Multilateralism

The EU 2030 Biodiversity Strategy had not yet been adopted when this chapter was completed. The European Commission failed to present its Communication on Safeguarding Nature by April 2020, as promised.[60] However, taking into account the statements made by the EU in the CBD meetings and pre-conferences, we dare to make the following recommendations on the basic objectives to be incorporated in the future strategy.

In the new global scenario of the coronavirus pandemic, the EU should align the objectives of its new strategy with those identified at the UN pre-conferences held before the Kunming Summit. The EU should also renew its commitment to further achieve its own Biodiversity Strategy to 2020, since its implementation at a domestic level has not achieved the desired level of success, as the European Environment Agency has reported. Furthermore, a renewed strategy should be valued positively because it will consolidate biodiversity as one of the goals of the EU environmental policy and, in particular, the EU biodiversity diplomacy based on the previous Biodiversity Strategy to 2020 and the Seventh Environment Programme.[61]

Continuity, coherence, and legitimacy derived from supporting the goals set by UN and UNEP are some of the features that characterize Green Diplomacy and also inform the EU diplomacy for biodiversity. The EU has always supported green multilateralism in alignment with the UN and UNEP proposals, at first as a recipient of international agreements and later taking on a more relevant role, proposing further goals, legal instruments, and global policies. Thus, as a normative power, the EU has been at the origin of the proposal and adoption of the Nagoya Protocol and has been one of the supporters of the global policies that are now rooted in the Aichi Targets and the Sustainable Development Goals, using them as "an overarching framework for policy development in the next 10 years [that] could provide an important step towards realising Europe's 2050 vision".[62] The support for these multilateral initiatives is at the core of the legitimacy of the EU external strategies, first because the EU cannot achieve its goals in isolation[63] and second because "most of the world's biodiversity is found outside the EU".[64]

Thus, the new EU 2030 Biodiversity Strategy should adopt an ambitious goal to update the Aichi Targets, assuming as a worldwide aim the restoration of 30% of terrestrial and marine ecosystems.[65] At the EU level, this goal will demand the extension of the Natura 2000 Network and a new cartography of nature in Europe which will be constructed through spatial planning covering marine and terrestrial areas and, for the first time, urban areas that represent a threat in terms of their contribution to climate change and habitat degradation. The link between climate change and biodiversity is recognized in the design of forestry measures, ranging from preserving and mapping primal forests to the planting of new ones as CO_2 sinks.

The CBD CoP 15 scheduled to be held in Kunming, China in October and November 2020 has as its main goal the adoption of a strategy for the post-2020 international biodiversity framework that can be considered a "new deal for nature". Preparations for this meeting were launched in late 2018, although unfortunately, the pre-conferences leading to CBD CoP15 have been postponed due to the coronavirus crisis. Nonetheless, at the previous meetings of the biodiversity governance institutions, some of the goals of the new Strategy for Post-2020 were already established. As happened with the 2012 Strategy and with the Aichi Targets, many goals will be recovered and will be repeated in the next strategy. Those targets will be a horizon of legal expectations – a contingent contract that will inform domestic and global policies leading states to adopt internal measures according to their national interests and possibilities. The position to be adopted by the EU and its member states has been shaped by last year's performance in the field of biodiversity protection in accordance with the legal procedures involving consultation with interested stakeholders.[66]

The European Parliament, which increased its external competences in the Lisbon Treaty, has adopted a resolution compelling other European institutions and the member states to assume obligations and binding goals post-2020 to protect global biodiversity from the threats of loss of species and ecosystems and climate change.[67] The European Parliament has been calling for ambitious and binding targets, monitoring mechanisms, detailed timetables, and performance indicators to fight for the preservation of global biodiversity as a priority. Pushing forward the Aichi Targets and the Strategy goals and their indicators, the European Parliament urges the EU to call on all parties to protect half the planet by 2050. It also insists that the EU lead these international negotiations by example and this is also our last recommendation.[68]

5. Final remarks

According to Elizabeth Maruma Mrema, the acting executive secretary of the CBD, "humanity will have given up on planet Earth if world leaders cannot reach an agreement this year to stop the mass extinction of wildlife and destruction of life-supporting ecosystems".[69] The EU wants to lead the first meeting of global leaders before the CBD and has declared to the UN General Assembly that the EU and its member states "remain committed to working with all our partners towards a successful and impactful Summit before the CBD Conference of the Parties that will agree the post 2020 Global Biodiversity Framework".[70]

The end of 2020 will be the right moment to assess the efficiency of EU diplomacy for biodiversity by weighing the achievements and failures recorded at the different conferences on biodiversity that will take place this year. Against the dramatic backdrop of the coronavirus crisis, the EU is committed more than ever to protecting biodiversity worldwide. In these times of the pandemic, biodiversity may be threatened by new risks to add to pollution, species loss, and climate change. The new situation should inspire a more coherent international wildlife law regime that is not based only on the regulation of the legal trade of endangered species, but also on the preservation of nature and its protection against human interference that endangers the fate of the planet.

Notes

1 As stated by the EU representative at the 13th meeting of the Conference of the Parties to the Convention on the Conservation of Migratory Species of Wild Animals (CMS COP 13) (Gandhinagar, India, 15–22 February 2020)- Statements by the EU and its Member States, 6320/20, 10 March 2020, p. 2.
2 UNTS, vol. 1760, p. 79.
3 See European Commission, Roadmap, EU Biodiversity Strategy to 2030, 23 December 2019 and the calendar for the adoption of the future Communication Safeguarding Nature, available from: https://ec.europa.eu/info/law/better-regulation/have-your-say/initiatives/12096-EU-2030-Biodiversity-Strategy [accessed 2 May 2020].
4 Morgera, 2012a.
5 The next EU Strategy for Biodiversity will be adopted at the end of May 2020 and it will incorporate the fight againts the pandemic as a new internal and global goal. See Bulletin Quotidien Europe N° 12485, 13 May 2020, available from: https://agenceurope.eu/en/bulletin/sommaire/12485 [accessed 13 May 2020].
6 CBD Open-Ended Working Group, "Zero draft text" of the Convention on Biological Diversity post-2020 global biodiversity framework, Doc. CBD/WG2020/2/3, 6 January 2020, available from: www.cbd.int/doc/c/efb0/1f84/a892b98d2982a829962b6371/wg2020-02-03-en.pdf [accessed 2 May 2020].

7   See the 13th meeting of the Conference of the Parties to the Convention on the Conservation of Migratory Species of Wild Animals (CMS CoP 13), Statements by the EU and its Member States, Doc. 6320/20, 2020.

8   EU priorities at the United Nations and the 74th United Nations General Assembly (September 2019 – September 2020), Doc. 10895/19, 15 July 2019.

9   European Commission Communication: The European Green Deal, COM(2019)640 final, 11.12.2019, p. 13.

10   Chasek, 2001, p. 1.

11   European Commission Communication: Our life insurance, our natural capital: an EU biodiversity strategy to 2020, COM(2011)244 final, 3.05.2011. This strategy has 6 thematic targets and 20 actions. The Target 6 is 'Helping stop the loss of global biodiversity.

12   EU 6th National Report on the Convention on Biological Diversity presented to the Clearing-House Mechanism of the Convention on Biological Diversity.

13   Fajardo, 2017.

14   Decision No 1386/2013/EU of the European Parliament and of the Council of 20 November 2013 on a General Union Environment Action Programme to 2020 'Living well, within the limits of our planet', OJ No. L 354/171, 28.12.2013, p. 176.

15   Directive 2009/147/EC of the European Parliament and of the Council of 30 November 2009 on the conservation of wild birds, OJ L No. 20, 26.1.2010 and Council Directive 92/43/EEC of 21 May 1992 on the conservation of natural habitats and of wild fauna and flora, OJ L No. 206, 22.7.1992.

16   Aichi Biodiversity Target 2 states that by 2020, ecosystem and biodiversity values need to be integrated into national and local planning, development processes, poverty reduction strategies and accounts.

17   A/RES/70/1.

18   Börzel and Buzogány, 2019.

19   European Green Deal, COM(2019)640 final, p. 7.

20   European Environment Agency, December 2019, Member States' reporting under the Birds and Habitats Directive. See V. Sazatornil et al., 2019.

21   European Environment Agency, 2019, The European environment – state and outlook 2020. Knowledge for transition to a sustainable Europe, (EEA 2019, hereinafter).

22   *Ibid.*

23   Thus, an IIED Issue Paper 2019 says, "Since the 1970s, formal protected area coverage has increased 660 per cent. However, the global populations of most major animal groups have declined by roughly 60 per cent. Simply declaring 'parks' isn't enough to halt biodiversity decline. Indigenous people and local communities own around 25 per cent of the world's land area, and they need support, in terms of tenure rights, resources and economic opportunities, that help them steward biodiversity. Beyond protected areas other mechanisms include paying for conservation services, with jobs as well as direct payments and supporting biodiversity friendly small-holder production such as agroforestry" p. 5.

24   European Parliament Resolution of 15 January 2020 on the European Green Deal (2019/2956(RSP).

25   European Green Deal, COM(2019)640 final.

26   Fajardo, 2017.

27   Art. 191.1 Treaty on the Functioning of the European Union (TFEU).

28   Since the EU competence should be exercise "without prejudice to Member States' competence to negotiate in international bodies and to conclude international agreements", article 191.4 TFEU.

29   Morgera, 2012b; Marín Durán and Morgera, 2012; Vedder, 2010.

30   So flexible solutions must be considered not just to convince member states, but also to reassure third countries of the EU's competence and commitments on the issues under

discussion in international negotiations. Thus, biodiversity protection in Antarctica has been extended through the creation of protected marine areas in a complex process in which the EU and member states have been drivers of the change, when marine biodiversity required a change of approach from sustainable exploitation of fisheries to plain conservation, Fajardo, 2019.

31 Council Decision of 25 October 1993 concerning the conclusion of the Convention on Biological Diversity, OJ No. L 309, 13.12.1993.
32 *Ibid.*
33 COM(2011)244 final.
34 Kramer, 2012.
35 Para. 11 of the 7th Environmental Action Program.
36 *Ibid.*, para 102, 7EAP.
37 See article. 3.2. TFEU.
38 See EU 6th National Report on the Convention on Biological Diversity presented to the Clearing-House Mechanism of the Convention on Biological Diversity, p. 51.
39 The Green Diplomacy Network (GDN hereinafter) was created in 2003 as an intergovernmental tool of the EU member states to promote EU environmental interests both in international organizations where the EU had not granted the status of member, but also for the member states to conduct multilateral and bilateral relations besides the EU, Fajardo, 2010.
40 Nicolson, 1950, p. 234.
41 These four actions are:

Action 17: Reduce the impacts of EU consumption patterns on biodiversity and make sure that the EU initiative on resource efficiency, our trade negotiations and market signals all reflect this objective.
Action 18: Target more EU funding towards global biodiversity and make this funding more effective.
Action 19: Systematically screen EU action for development cooperation to reduce any negative impacts on biodiversity.
Action 20: Make sure that the benefits of nature's genetic resources are shared fairly and equitably.

42 New Consensus on Development 'Our World, Our Dignity, Our Future', 2017, para. 44.
43 Marín and Morgera, 2012.
44 7th Environmental Action program, para. 22.
45 See Chapter 4.
46 13th meeting of the Conference of the Parties to the Convention on the Conservation of Migratory Species of Wild Animals (CMS COP 13) (Gandhinagar, India, 15–22 February 2020)- Statements by the EU and its Member States, 6320/20, 10 March 2020, p. 3, 7.
47 *Ibid.*
48 Manners, 2002; Bretherton and Vogler, 2006; Zielonka, 2008.
49 Morgera, 2012b, p. 380.
50 Marín Durán and Morgera, 2012.
51 See the EU agreements and Political Dialogue with Indonesia (Art. 27.3c) and the Central American countries (Art. 38.1), Framework Agreement on comprehensive partnership and cooperation between the European Community and its member states, of the one part, and the Republic of Indonesia, of the other part, OJ No. L 125, 26.04.2014, and the Political Dialogue and Cooperation Agreement between the European Community and its member states, of the one part, and the Republics of Costa Rica, El Salvador, Guatemala, Honduras, Nicaragua and Panama, of the other part, OJ No. L 111, 15.4.2014 (Central America Political Dialogue and Cooperation Agreement hereinafter).

52 See the numerous documents on the EU and its member states on the CBD CoP as well as in the CoPs of its protocols, Doc. 15700/18.
53 Central America Political Dialogue and Cooperation Agreement, article 38.2.c and e.
54 Büge et al., 2015.
55 COM(2011)244 final, Target 6. Action17.
56 COM(2011)244 final, Target 6, Actions 17 and 19.
57 Lebzelter and Marx, 2020, p. 1.
58 High Representative of the Union for Foreign Affairs and Security Policy, Joint Report to the European Parliament and the Council, Report on the Generalised Scheme of Preferences covering the period 2018–2019, JOIN(2020)3 final, 10.2.2020, p. 9.
59 Thus, the impact assessment on the Free Trade Agreement between the European Union and its member states and the Republic of Korea says that there are no indications of "any observable effect of the EU-Korea FTA" on "environmental areas, such as air pollution, water quality, biodiversity, waste management and deforestation", Commission Staff Working Document Evaluation of the Implementation of the Free Trade Agreement between the European Union and its Member States, of the one part, and the Republic of Korea, of the other part, SWD(2019) 102 final, 7.3.2019, p. 42.
60 After a first postponement, the adoption of the Communication was deferred until 29 April, See *Bulletin Quotidien Europe* N°12454, 26 March 2020 and EUROPE B12450A31.
61 OJ No. L 354, par. 11 and 23 of the preamble.
62 European Environment Agency, December 2019, Member States' reporting under the Birds and Habitats Directive, p. 16.
63 *Ibid.*
64 EU 6th National Report on the Convention on Biological Diversity presented to the Clearing-House Mechanism of the Convention on Biological Diversity, available at the CBD National Report Information Portal, available from: www.cbd.int/nr6/ [accessed 15 May 2020].
65 The previous strategy, the EU 2020 Biodiversity Strategy and the 7th Environment Programme only targeted to restore 15% of degraded ecosystems, EC (2019) Natural Capital Accounting: Overview and Progress in the European Union, p. 11.
66 European Commission, Roadmap, EU Biodiversity Strategy to 2030.
67 European Parliament Resolution of 16 January 2020 on the 15th meeting of the Conference of Parties (COP15) to the Convention on Biological Diversity, 2019/2824(RSP).
68 European Parliament Resolution of 15 January 2020 on the European Green Deal (2019/2956(RSP).
69 UNEP, Biodiversity in grave danger: what can be done in 2020? 28 January 2020, available from: www.unenvironment.org/news-and-stories/story/biodiversity-grave-danger-what-can-be-done-2020 [accessed 10 March 2020].
70 EU Explanation of Position – United Nations General Assembly: Adoption of the Resolution on the Scope, format, modalities and organization of the UN Summit on Biodiversity, 1 April 2020.

References

Börzel, T.A. and Buzogány, A., 2019. Compliance with EU Environmental Law. The Iceberg Is Melting. *Journal of Environmental Politics*, 28, 315–341.

Bretherton, C. and Vogler, J., 2006. *The European Union as a Global Actor*. New York: Routledge.

Büge, M., Meijer, K. and Wittmer, H., 2015. *International Financial Instruments for Biodiversity Conservation in Developing Countries – Financial Mechanisms and Enabling Policies for Forest Biodiversity*. Background paper for the European Report on Development.

Chasek, Pamela S., 2001. *Earth Negotiations. Analyzing Thirty Years of Environmental Diplomacy*. New York: United Nations University Press.

Fajardo, T., 2017. La Política Exterior de la Unión Europea en materia de Medio ambiente. In Beneyto Pérez, J.M. (Dir.), *Tratado de Derecho y Políticas de la Unión Europea, Tomo IX, Acción Exterior de la Unión Europea*. Madrid: Thomson-Reuters Aranzadi, 303–379.

IPBES, 2019. *Global Assessment on Biodiversity and Ecosystem Services*. Available from: https://ipbes.net/news/ipbes-global-assessment-preview.

Krämer, L., 2012. *EU Environmental Law*. London: Sweet & Maxwell.

Lebzelter, T. and Marx, A., 2020. Is EU GSP+ Fostering Good Governance? Results from a New GSP+ Compliance Index. *Journal of World Trade*, 54(1), 1–30.

Manners, I., 2002. Normative Power Europe: A Contradiction in Terms? *Journal of Common Market Studies*, 40, 235–258.

Marín Durán, G. and Morgera, E., 2012. *Environmental Integration in the EU's External Relations. Beyond Multilateral Dimensions*. Oxford: Hart Publishings.

Morgera, E. (Editor), 2012a. *The External Environmental Policy of the European Union. EU and International Law Perspectives*. Cambridge: Cambridge University Press.

Morgera, E., 2012b. The Trajectory of EU Biodiversity Cooperation: Supporting Environmental Multilateralism Through EU External Action. In Morgera, E. (Editor), *The External Environmental Policy of the European Union. EU and International Law Perspectives*. Cambridge: Cambridge University Press, 235–259.

Nicolson, H., 1950. *Diplomacy*, 2nd ed. Oxford: Oxford University Press.

Sazatornil, V., Trouwborst, A., Chapron, G., Rodríguez, A. and López-Bao, J.V., 2019. Top-Down Dilution of Conservation Commitments in Europe: An Example Using Breeding Site Protection for Wolves. *Biological Conservation*, 237, 185–190.

Vedder, H., 2010. The Treaty of Lisbon and European Environmental Law. *Journal of Environmental Law*, 22, 285–299.

Zielonka, J., 2008. Europe as a Global Actor: Empire by Example? *International Affairs*, 84, 471–484.

4   The role of the EU in the promotion of sustainable development through multilateral trade

Xavier Fernández-Pons

1. Introductory remarks

The European Union (EU), in exercise of its Common Commercial Policy, has tended to present itself as one of the members of the World Trade Organization (WTO) that is most concerned with promoting a more sustainable multilateral trading system.[1] Since the reform enacted by the Treaty of Lisbon in 2007, the Treaty on European Union (TEU) provides that the EU shall promote its values and interests "in its relations with the wider world", contributing to "the sustainable development of the Earth" and "free and fair trade".[2] The term "fair trade" is particularly ambiguous and can cover a wide range of content. However, it is often used to designate international trade that takes into account, among other aspects, social and environmental considerations in the production of goods and services.[3] In any event, these and other generic provisions of the TEU relating to the EU's external action[4] require the adoption of specific measures.

In October 2015, shortly after the United Nations (UN) adopted the Sustainable Development Goals (SDGs) under the 2030 Agenda,[5] the European Commission launched a new trade strategy with its Communication entitled "Trade for All".[6] Taking into account the growing criticism of certain aspects of the multilateral trading system and of some of the new bilateral or regional agreements then under negotiation, such as the Transatlantic Trade and Investment Partnership (TTIP) with the United States (US) under the presidency of Barack Obama, the then EU trade commissioner, Cecilia Malmström, underlined that "[i]n this new strategy . . . trade policy will become more responsible, meaning it will be more effective, more transparent and will not only project our interests, but also our values".[7] In particular, the European Commission envisaged various initiatives to ensure that European consumers "can be confident in the products they buy in a global economy", including "respect for human rights, labour rights and the environment in the way they are produced" and "a trade agenda to promote sustainable development, human rights and good governance".[8] The European Commission also pointed out the need for "reinvigorating the multilateral trading system", noting that "despite some improvements, it has remained largely unchanged for the past two decades".[9]

However, updating the multilateral trading system has proved to be an arduous task. The Doha Round and other global negotiations have highlighted the difficulties in reaching broad consensus between advanced and emerging economies in an increasingly multipolar world. For example, since 2014, the EU and other WTO members have been negotiating an Environmental Goods Agreement (EGA) to remove trade barriers on environmental or "green" goods (e.g., recycling machinery, wind turbines, and solar panels) without success.[10] Recently, the crisis of the multilateral trading system has become even more acute under the presidency of Donald Trump, with the trade wars he has launched, especially against China, and the US blockade on the renewal of the members of the WTO Appellate Body.[11]

In this complex context, the EU has continued to defend the institutional and legal bases for global free trade, but the EU has also stressed its calls for a reform of the WTO system, underlining, among other aspects, a need to promote fairer and more sustainable international trade and a need for the WTO to contribute more clearly to the pursuit of the SDGs, going far beyond the current Doha Round negotiations on "elimination of the most harmful fisheries subsidies" (SDG 14.6).[12]

The 2018 EU Trade Defence Report, presented on 27 March 2019, stresses the need for effective protection against "unfair trade", taking into account "social and environmental protection standards" in third countries.[13] The new president of the European Commission, Ursula von der Leyen, has provided in the Political Guidelines for 2019–2024, presented to the European Parliament on 16 July 2019, a "European Green Deal", and has insisted that the EU must strengthen its role as "global leader and standard setter through a strong, open and fair trade agenda" to "export our values across the world".[14] The president of the European Commission, in her mission letter of 10 September 2019 to the new commissioner for trade, Phil Hogan, also insists on the need to use "trade tools to support sustainable development".[15] In December 2019, the European Commission presented its Communication on "The European Green Deal", which envisages various measures to promote a stronger "green deal diplomacy", seeking to ensure that the EU's trade policy "serves as a platform to engage with trading partners on climate and environmental action".[16]

Within the framework outlined here, this chapter will present various initiatives that have been launched by the EU to meet these aims: the EU's Generalised Scheme of Preferences and its special incentive arrangement for sustainable development and good governance, addressed in section 2; the inclusion of a chapter on sustainable development in new regional trade agreements concluded by the EU with third countries, in section 3; the EU's promotion of sustainable public procurement, in section 4; the recent reform of the EU's regulation of anti-dumping measures, partly addressing so-called social and ecological dumping, in section 5; and some EU trade measures related to carbon footprint, such as the proposal to establish a carbon border tax contained in "The European Green Deal", in section 6. Some of these EU initiatives have been strongly challenged by certain emerging or developing countries, which may sometimes see them as new forms of disguised protectionism or eco-imperialism.[17] Lastly, in section 7, the chapter will seek to assess the EU's ability to promote sustainable development around the world through trade instruments, before concluding with some final remarks.

2. The EU's Generalised Scheme of Preferences (GSP) and its special incentive arrangement for sustainable development and good governance (GSP+)

As is well known, in 1979 the Contracting Parties of the General Agreement on Tariffs and Trade (GATT) 1947 approved the so-called Enabling Clause, which is still in force and allows for the establishment of "generalised" schemes of tariff preferences for products originating in developing countries.[18]

As for the beneficiary countries, the Enabling Clause provided, in principle, that preferences should be "generalised" for developing countries, excluding arbitrary selections. The Enabling Clause did explicitly contemplate that a more favourable treatment could be granted to least-developed countries (LDCs), but it did not clearly allude to the possibility of making gradations between different developing countries based on other criteria.

However, the EU has been establishing special incentive arrangements by granting specific preferences to developing countries that, without being LDCs, meet certain social or environmental conditions and requirements. Initially, Regulation 2501/2001/EC of 10 December 2001[19] established five different arrangements: 1) a general arrangement for developing countries in general; 2) a special incentive arrangement for the protection of labour rights; 3) a special incentive arrangement for the protection of the environment; 4) a special arrangement to combat drug production and trafficking; and 5) a special arrangement for LDCs, the so-called Everything But Arms (EBA) initiative.

At first, the compatibility of the various special incentive arrangements with the Enabling Clause was unclear, and India even filed a complaint at the WTO in March 2002, giving rise to the dispute *European Communities (EC) – Conditions for the Granting of Tariff Preferences to Developing Countries*,[20] in which the Appellate Body specified the criteria for determining the permissibility of possible incentive arrangements. According to the Appellate Body, Paragraph 3(c) of the Enabling Clause allows for certain distinctions to be made between developing countries as a function of their "development, financial and trade needs", which may vary from country to country. The Appellate Body considered that such "needs" should in any case be assessed according to "an objective standard",[21] concluding that "broad-based recognition of a particular need, set out in the WTO Agreement or in multilateral instruments adopted by international organizations, could serve as such a standard".[22]

Subsequently, the EU designed a new GSP, through Regulation 980/2005/EC of 27 June 2005,[23] trying to respect the criteria indicated by the Appellate Body and providing a "special incentive arrangement for sustainable development and good governance", known as GSP+. This entailed, to a large extent, the integration of three previously separate special incentive arrangements (concerning labour rights, the environment, and drugs) into a single new arrangement. The new design of the GSP+ has not been challenged in the WTO dispute settlement system.

The EU's current GSP, established by Regulation 978/2012/EU of 25 October 2012 and effective from 1 January 2014 for a period of ten years,[24] maintains substantially the same GSP+ arrangement, situated between the general arrangement granted to developing countries with a lower-middle income (Standard GSP) and the most favourable regime (EBA) granted to LDCs. The GSP+ arrangement is currently granted to lower-middle income countries that, according to Art. 9 of the aforementioned Regulation, are "vulnerable" ("due to a lack of diversification and insufficient integration within the international trading system") and that have ratified and are committed to effectively implementing various core international conventions on human and labour rights, the environment, and good governance listed in Annex VIII, including eight environmental conventions promoted by the UN.

Currently, according to the list on 1 January 2019, eight countries benefit from the GSP+: Armenia, Bolivia, Cape Verde, Kyrgyzstan, Mongolia, Pakistan, the Philippines, and Sri Lanka.[25] This is a rather small number compared to the beneficiaries of the Standard GSP (15) or the LDCs covered by the EBA (48), which are essentially designated on the basis of their *per capita* income and are not subject to so many conditions.[26]

An extensive report on the mid-term evaluation of the EU's current GSP in 2018, prepared by the consultancy Development Solutions for the European Commission, notes that the GSP+ has had positive economic and social impacts on beneficiary countries, but that the environmental impacts are less evident, regardless of their ratification of the corresponding international conventions.[27] It seems therefore that the GSP+ is, at present, a rather limited instrument for promoting effective environmental protection abroad and it is understandable that, in view of the future revision of the EU's current GSP which will expire on 31 December 2023, the European Parliament has underlined, in a non-legislative resolution of 14 March 2019, the need to place more emphasis on environmental requirements in an attempt to increase the EU's influence in this respect.[28]

3. The inclusion of a chapter on sustainable development in the new regional trade agreements (RTAs) concluded by the EU with third countries

Although the conclusion of preferential or regional trade agreements (RTAs) is not new, the aforementioned difficulties in the Doha Round and the current WTO crisis may have contributed, at least in part, to the fact that in recent years there has been a marked proliferation of regional and mega-regional trade negotiations. The EU has pushed forward with an ambitious agenda of negotiations with third countries and often refers to its most recent agreements as the "New Generation Free Trade Agreements".[29] These include, for example, agreements with: South Korea;[30] Colombia-Peru-Ecuador;[31] Central America;[32] Canada, known as the

Comprehensive Economic and Trade Agreement (CETA);[33] Singapore;[34] Japan;[35] Vietnam;[36] and Mercosur.[37] All these agreements are characterized by their broad scope and usually include provisions on trade in goods, services, intellectual property, foreign investment, government procurement, competition, and other issues that go beyond the provisions of the multilateral trading system.

It is interesting to note here that all these agreements contain a chapter dedicated to "Trade and Sustainable Development" (TSD chapter). This is fully in line with the aforementioned "Trade for All" and "European Green Deal" Communications and the Political Guidelines for 2019–2024, which insist that all RTAs concluded by the EU must incorporate a TSD chapter. In this respect, Ursula von der Leyen has stated: "I will ensure that every new [regional trade] agreement concluded will have a dedicated sustainable-development chapter and the highest standards of climate, environmental and labour protection, with a zero-tolerance policy on child labour".[38]

Taking as an example the EU-Mercosur agreement in principle announced on 28 June 2019, the relevant TSD chapter is comprised of 18 articles with substantive provisions on: objectives and scope (alluding to the SDGs), the right to regulate and levels of protection (establishing, for instance, that a party "should not weaken the levels of protection afforded in domestic environmental or labour law with the intention of encouraging trade or investment"), transparency, multilateral labour standards and agreements, multilateral environmental agreements, trade and climate change (with specific references to the Paris Agreement), trade and biodiversity, trade and sustainable management of forests, trade and sustainable management of fisheries and aquaculture, scientific and technical information (with an explicit reference to the precautionary principle), trade and responsible management of supply chains, etc. From an institutional point of view, it provides for the creation of a sub-committee on trade and sustainable development (TSD Sub-Committee) and a specific mechanism for dispute resolution.[39]

The systematic inclusion of TSD chapters in the RTAs concluded by the EU is a clear way to assert the EU's regulatory power at the international level, as some authors have already analysed,[40] and this is commendable. In contrast to the traditional focus of the multilateral trading system on trade liberalization and its scarce references to social and environmental issues, which are mainly contemplated by means of succinct and vague exceptions, the new RTAs promoted by the EU take a more holistic view, including explicit and multiple normative connections between economic, social, and environmental aspects. With the TSD chapters, the EU is promoting an international regulatory model that seeks to overcome the so-called fragmentation between various international legal regimes, which have tended to develop in isolation and with few interconnections between them. Ideally, the multilateral trading system could follow the EU's example in the future and incorporate a TSD chapter or agreement, but the current global context does not appear conducive to such a step forward.

In any case, the TSD chapters should not remain mere regulatory provisions and their effective application in practice must be ensured. Some analysts have warned, for example, that TSD provisions such as the ones contained in the aforementioned EU-Mercosur agreement could have very little practical relevance in the face of economistic policies such as those of President Jair Bolsonaro of Brazil.[41] Some authors from Mercosur countries have already criticized the pertinent TSD chapter, viewing it as a sign of a supposed new "regulatory protectionism" from the EU, which wants to "dye everything green".[42] On the other hand, it is worrying that the TSD chapters provide their own dispute settlement mechanism, which is softer than the mechanisms established to guarantee other chapters of RTAs.

In an attempt to evaluate the effectiveness of TSD chapters, the EU has included specific monitoring of such chapters within its periodic follow-up of the implementation of its RTAs. Although it is still too early to appreciate the full effectiveness of such chapters, in February 2018 the European Commission identified "15 action points" to make TSD chapters more effective.[43] These 15 points are organized under four headings (working together, enabling civil society, delivering, and communicating and transparency) and they include, for example, "ensuring that countries comply with their commitments through more assertive enforcement" (although no sanctions are envisaged), "facilitating the monitoring role of civil society", and "making EU resources available to support the implementation" of TSD chapters, especially targeted at developing countries, so that such chapters are not seen as rigid impositions from advanced economies on countries with very different levels of development and resources.[44]

4. The EU and sustainable public procurement

According to the WTO, public procurement "accounts for 10–15 per cent" of the gross domestic product (GDP) of a country on average, constituting a "significant market and an important aspect of international trade".[45] When the Marrakesh Agreements created the WTO, they also included a Government Procurement Agreement (GPA 1994) in an effort to promote transparency and non-discrimination in the public procurement of goods, services, and construction works. This agreement did not enter into the WTO multilateral agreements, which bind, as a single undertaking, all WTO members, but it was included as one of the so-called "plurilateral" agreements of the WTO, which WTO members may or may not accept. The GPA 1994 was mainly accepted by advanced economies, because emerging and developing countries often prefer not to limit their sovereignty on an issue as sensitive as government procurement. This particular feature of the GPA 1994, as a plurilateral agreement, may have been one of the factors that facilitated its updating, which is much more difficult with multilateral agreements. Thus, a new revised GPA was adopted in 2012 and entered into force on 6 April 2014.[46] Currently, the GPA 2012 consists of 20 parties covering 48 WTO members (counting the EU and its 27 member states after Brexit).[47]

Among the improvements introduced in the GPA 2012, there is a "more explicit recognition of the right of procuring entities to promote environmental values and sustainability".[48] It should be noted that the EU and the US under the Obama administration were decisive in the negotiation of the GPA 2012, as none of the so-called BRICS countries (Brazil, Russia, India, China, and South Africa) is a party to the agreement, either in its original or revised version. Therefore, it was possible to agree on a text that essentially responds to the most commonly held views of advanced economies, with the EU and its member states among the main drivers of sustainable public procurement.[49]

The GPA 2012 includes various provisions specifically related to the environment. Thus, for example, Art. X on "Technical Specifications and Tender Documentation" specifies: "For greater certainty, a Party, including its procuring entities, may, in accordance with this Article, prepare, adopt or apply technical specifications to promote the conservation of natural resources or protect the environment".[50]

Furthermore, Art. XXII:8(a) GPA 2012 provides that the parties "shall adopt and periodically review a work programme, including a work programme on sustainable procurement". In its decision of 30 March 2012, the WTO Committee on Government Procurement approved a "Work Programme on Sustainable Procurement", which provides for the examination of several topics, including the ways in which "the concept of sustainable procurement is integrated into national and sub-national procurement policies".[51]

Since the adoption of the GPA 2012, the EU has also updated its rules in this area. In particular, its current Directive 2014/24/EU of 26 February 2014 on public procurement[52] contains numerous references to sustainable development and to specific social and environmental aspects, such as life-cycle analysis. Of course, the effectiveness of those EU provisions depends, to a large extent, on their proper transposition and implementation in the member states.[53]

Good practices on sustainable public procurement in the EU can serve as an example for other countries and be an important benchmark for the "Work Programme on Sustainable Procurement" that is now being promoted under the GPA 2012, as demonstrated by the symposium on sustainable public procurement organized by the WTO in Geneva on 22 February 2017.[54] It should be noted, however, that Donald Trump has very recently announced that the US could withdraw from the GPA 2012,[55] a circumstance that would greatly weaken the regulation of public procurement at the WTO.

5. The recent reform of the EU regulation of anti-dumping measures and the issue of so-called social and ecological dumping

WTO law allows its members to adopt various trade defence measures, including anti-dumping measures. Based on the provisions of Art. VI GATT 1994, the current WTO Anti-Dumping (AD) Agreement defines what is understood by dumping on a global level and regulates how WTO members injured by such practices can adopt measures to counteract it. In accordance with these multilateral provisions, the EU, like other WTO members, has its own regulations for imposing anti-dumping measures, which are revised successively. The current EU basic regulation is contained in Regulation 2016/1036/EU of 8 June 2016,[56] which has been substantially amended by Regulation 2017/2321/EU of 12 December 2017[57] and Regulation 2018/825/EU of 30 May 2018,[58] introducing a new methodology for determining the existence of dumping. Henceforth, references will be made to the latest consolidated version of the basic anti-dumping Regulation (the codified basic AD Regulation) of 8 June 2018.[59]

From the specific perspective of this chapter, the new EU anti-dumping methodology is interesting because, as will be detailed later, it takes into account, at least partially, so-called social dumping and ecological dumping, which are notions that have not traditionally been covered within the multilateral legal definition of dumping. In order to appreciate the scope of the new methodology established by the EU, it is useful to remember what is understood by dumping in the framework of the multilateral trading system and the context in which the EU has approved its new methodology.

In the multilateral trading system, dumping is deemed to occur when, according to the current provisions of Art. VI:1 GATT 1994 and Art. 2 AD Agreement, "products of one country are introduced into the commerce of another country at less than the normal value of the products". The key issue is how to calculate the "normal value". The aforementioned WTO rules provide several formulas, but the general rule is that the normal value must be based on the prices and costs of production of the product in the country of origin. Thus, for example, a very clear case of dumping occurs when a company exports its products at a price below its cost of production, in accounting terms. However, if a company exports its products at low prices and its prices and costs at origin are equally low, it is not dumping. On this basis, it is understandable that the AD Agreement is not typically considered to contemplate the phenomena that some authors have been describing as social dumping and ecological dumping, i.e., exporting at low prices thanks to lax social and environmental conditions in the countries of origin, which consequently permit a reduction in production costs.[60]

The multilateral trading system has traditionally allowed its members to apply a special methodology for calculating dumping in the case of products exported from non-market economy (NME) countries, given that their domestic prices and costs may not be an appropriate benchmark in that they are highly conditioned by strong government intervention in the economy.[61] Thus, in order to defend themselves against exports from NMEs, many importing countries have calculated their normal value on the basis of prices and costs in a surrogate third country with a market economy. For example, for a long time, products originating in China have been subject to numerous anti-dumping measures imposed by the EU, the US, and other countries with relative ease by calculating the normal value of such products according to the prices and costs of other third countries with a market economy (such as, for instance, Brazil, Colombia, or Thailand).[62]

When China joined the WTO in 2001, the Protocol of Accession explicitly provided that China could be considered, initially, as an NME, but it was contemplated (on the assumption that the country would continue with reforms of its economic system) that this specific treatment would expire within 15 years.[63] At the expiry of the deadline in December 2016, China interpreted that it should automatically be treated as a market economy, requiring the application of the general methodology for the calculation of dumping, based on prices and costs in the Chinese market itself. However, other WTO members, such as the EU and the US, disagree with this Chinese interpretation and consider that China has many reforms pending and can still be treated as an NME.[64] In response, China has filed complaints against the EU and the US before the WTO for continuing to treat it as an NME in their application of anti-dumping measures, but no final report on this thorny issue has yet been circulated by a panel or by the Appellate Body.[65]

This was precisely the context in which the EU decided to undertake a reform of its own methodology to determine the existence of dumping and calculate the dumping margin. The special regime previously provided for NME countries has been replaced by a new special regime for countries with "significant distortions".[66] Now, when the EU considers that a country has "significant distortions", either in its market as a whole or in a specific sector, the EU can determine the normal value of the exported product without taking into account the domestic prices and costs in the exporting country, but instead taking into account, for example, the prices and costs of a surrogate third country without "significant distortions", but with a similar level of economic development, and giving preference, where appropriate, "to countries with an adequate level of social and environmental protection".[67] This latter provision is one of the precepts that introduce social and environmental aspects into the codified basic AD Regulation, but it is not the only one.

The concept of "significant distortions" is defined as "those distortions which occur when reported prices or costs, including the costs of raw materials and energy, are not the result of free market forces because they are affected by substantial government intervention", taking into account certain elements such as, for instance, "public policies or measures discriminating in favour of domestic suppliers or otherwise influencing free market forces" and "wage costs being distorted".[68]

Moreover, when the amount of the anti-dumping duty is not calculated on the basis of the dumping margin practiced, but on the basis of the injury margin (which is a lower margin, intended only to remove the injury to the EU industry), the EU will take into account "the actual cost of production of the Union industry, which results from multilateral environmental agreements, and protocols thereunder, to which the Union is a party, or from International Labour Organization (ILO) Conventions listed in Annex IA to this Regulation".[69]

These new and modernized EU rules on anti-dumping measures have been heavily criticized by several WTO members, particularly by China. China considers that, although the reform has been formulated in general terms and the concept of "significant distortions" can theoretically be applied to many different countries, it is a reform mainly aimed at continuing to facilitate the EU's imposition of anti-dumping measures against exports of Chinese products, circumventing the debated question of whether China can currently be considered an NME. China stresses that although the existence of "significant distortions" will be assessed on a case-by-case basis, the codified basic AD Regulation also provides that the

European Commission can prepare guidance reports on "significant distortions" identified in a given country, and the European Commission has indeed started to do so precisely by making an extensive report on China.[70] The Ministry of Commerce of China (MOFCOM) believes that many of the new provisions of the EU's codified basic AD Regulation are incompatible with the multilateral trading system and its AD Agreement, claiming that "WTO rules do not include the concept of 'significant market distortions' and the rule of social and environmental dumping" and reserving the possibility of filing a complaint through the WTO dispute settlement system.[71]

In the framework of the WTO Committee on Anti-Dumping Practices, there has been much discussion about the new EU rules. In addition to harsh criticism from China, other countries have expressed their disappointment. Russia, which is the second country on which the European Commission is preparing a report on its "significant distortions",[72] has stated that "the amendments to the EU's basic Anti-Dumping Regulation were inconsistent with the WTO rules".[73] Saudi Arabia, Bahrain, and Oman have expressed concerns about the new method to determine normal value in the case of "significant distortions" of the market in the exporting country.[74] Argentina holds the view that "the EU's new rules could not find textual support in the WTO covered agreements and would violate the objectives of the AD Agreement".[75] In contrast, it is significant that the US has given some support to the EU, underlining the need to have "appropriate tools" to address "certain distortions that impacted international trade".[76]

Despite the criticism from China and other countries, no claims against the new provisions of the EU's basic AD Regulation as such have been presented through the WTO dispute settlement system thus far. It remains to be seen how the new EU rules will be applied to specific cases in practice, which is still in the early stages. Furthermore, although the new EU rules are controversial, they may well be overshadowed by the drastic unilateral trade measures taken against China by the Trump administration.

In terms of doctrine, some legal scholars downplay the novelties introduced in the EU's basic AD Regulation, observing that most of the examples of detailed "significant distortions" refer to government interventions in the market traditionally carried out by NMEs or countries with an economic regime in transition and do not relate to the protection of certain social or environmental standards, so that the novelties would essentially amount to new terminology or legal coverage to carry on as usual imposing, with relative ease, anti-dumping measures against exports of Chinese products, and basically putting "old wine in a new bottle".[77]

By contrast, other legal scholars value these new EU rules positively, highlighting the indicated interconnections between trade, social, and environmental rules,[78] or stressing that the EU is equipping itself with trade defence instruments that address issues of social and ecological dumping for the first time, although timidly and overlapping with other objectives.[79] Still other legal scholars, while supporting this basic philosophy, observe that the new EU anti-dumping methodology and some of the new terms, such as "significant distortions" of the market, cannot easily fit within the current text of the AD Agreement.[80]

Certainly, an eminently literal interpretation of the WTO rules could hardly coincide with the new EU anti-dumping provisions. However, it is worth remembering that, according to the customary rules for the interpretation of international treaties codified in the Vienna Convention on the Law of Treaties (VCLT) and explicitly applicable to the interpretation of WTO agreements,[81] it is also necessary when interpreting treaties to take into account, together with the current meaning of their terms, their "context", their "object and purpose" and, among other elements, "any relevant rules of international law applicable in the relations between the parties", such as those related to sustainable development, which allow for evolving interpretations.[82]

From the EU's point of view, its modernized regulation of anti-dumping measures should ideally serve as a model for any future reform of trade defence measures in the WTO. The latest annual report of the European Commission on its trade defence instruments, in which it defends the EU's recent modernization, regrets that the text of the corresponding WTO rules have not been updated since the Marrakesh Agreements, taking the view that the "new global market realities" and the idea that "an open trade policy can only be built on sustainable trade that respects a minimum number of shared values" should be taken into account.[83]

However, the prominence of the countries that have been critical of the new EU provisions on anti-dumping is an indication of the difficulties that confront the EU in acting as a global regulatory leader in the process of devising new global rules on this controversial issue.

6. EU's trade measures related to carbon footprint

Under the multilateral rules of the WTO, it is difficult to justify the imposition of trade restrictions on products based on processes and production methods (PPMs) that do not leave a physical trace in the final product, which are usually called non-product-related PPMs (npr-PPMs).[84] However, such justification is possible under certain conditions, as already shown by the Appellate Body jurisprudence in the landmark *US – Shrimp* case.[85] Taking this WTO case law into account, the EU has been imposing various restrictions on trade based on npr-PPMs, such as, for example, on the import of fish or timber in terms of the sustainable management of fisheries and forests, as analysed in other chapters of this book.[86]

Here, we will briefly refer to another type of trade restriction which the EU has been promoting based on npr-PPMs and which is particularly relevant for a crucial environmental issue like the fight against climate change: measures that restrict or penalize the marketing of certain products based on their carbon footprint, taking into account carbon emissions during their production process in the country of origin. This is an issue on which the WTO adjudicative bodies have not yet made an explicit pronouncement,[87] although two highly interesting disputes have already been brought against the EU over measures of this kind.

A first example arose in 2013, when Argentina submitted a request for consultations to the EU regarding Directive 2009/28/EC of 23 April 2009, known as the Renewable Energy Directive (RED) I,[88] and other complementary and transposition rules, claiming that they contained certain provisions that were incompatible with WTO rules.[89] Argentina argued that the distinction between sustainable and non-sustainable biofuels established in these EU rules was unjustifiably discriminatory against soybean biodiesel, which is a product of major importance in the Argentine economy and which was not, in principle, categorized as a sustainable biofuel. In contrast, other biodiesels (such as rapeseed biodiesel, which has significant production in the EU) were considered sustainable. Argentina noted that, under the EU rules, only sustainable biofuels would be considered for mandatory targets and could benefit from tax reductions and other support measures, providing that their sustainability took into account their entire life cycle, including their production process in the country of origin.[90] The EU rules initially provided that only those biofuels that had a net reduction in greenhouse gas emissions of 35% compared to those generated by a traditional fossil fuel deserved to be categorized as sustainable biofuels, and Argentina considered that this threshold was arbitrary and unduly penalized soybean biodiesel.[91] According to the default values calculated by the EU itself, soybean biodiesel was not sustainable (the net reduction was only 31%), whereas rapeseed biodiesel was just above the sustainability threshold (the net reduction was 38%).[92] It should be noted that Argentina did not question the general logic of the EU measures, but did question the threshold and the default values calculated for soybean biodiesel. In any case, the dispute was put on hold in the WTO dispute settlement system, without any report or mutually agreed solution.[93]

A second and more recent example concerns the complaint filed in December 2019 by Indonesia against the EU on certain measures concerning palm oil and biofuels based on palm crops.[94] Indonesia is the world's largest producer of palm oil (used, among other things, in biofuel production) and some studies indicate that the expansion of palm crops in Indonesia (and other tropical countries) has been taking place largely through the deforestation of primary forests, destroying important carbon sinks.[95] Indonesia notes that the EU's classification of biofuels as sustainable initially focused on emissions from direct land use, but that more recently, the EU has also sought to address the risk posed by the effects of so-called indirect land use change (ILUC), and that this would unduly penalize palm oil biofuel.[96] Thus, Indonesia claims that this biofuel has been heavily penalized by the new Directive 2018/2001 of 11 December 2018, known as the Renewable Energy Directive (RED) II,[97] the Delegated Regulation 2019/807/EU of 13 March 2019,[98] and other complementary rules. According to these new EU rules, palm cultivation is generally considered to present a high risk of ILUC, expanding at the expense of land that previously had large carbon sinks. Although there is a procedure to certify that a certain palm crop is sustainable, Indonesia claims that these EU measures generally penalizing palm oil are incompatible with WTO rules. Indonesia questions their scientific basis, claiming that "ILUC emissions cannot be measured with the level of precision required to be included in the EU's GHG emission calculation methodology",[99] and considers that the generic penalization of palm oil is not well founded, that the certification procedure envisaged is very complex and difficult to obtain in practice, and that the EU has not taken due account of the fact that Indonesia and other palm oil producers are developing countries.[100] This is a case of great economic and systemic interest. Pending any response that the WTO adjudicative bodies may give, it is significant that several countries (Argentina, Colombia, Costa Rica, Guatemala, Malaysia, and Thailand) have requested to join the consultations, generally claiming that they also cultivate palm and export its derivatives (oil or biofuel) to the EU.[101]

Finally, we will allude to one of the most prominent proposals in the aforementioned Political Guidelines for 2019–2024 and "The European Green Deal" Communication: the establishment of a "Carbon Border Tax to avoid carbon leakage", which "should be fully compliant with" WTO rules.[102]

Although it remains to be seen how this carbon border tax will be designed by the EU,[103] the logic of such a tax consists of taxing certain imported products, generally at the border or customs, according to the carbon emissions resulting from their production process at origin. Thus, for example, steel imported from a country where it is produced with very low carbon emissions and steel imported from another country where it is produced with high carbon emissions would be subject at customs, despite being physically similar or even identical products, to different carbon taxes, depending on their respective carbon footprint.

There has been much discussion in the literature as to whether or not such measures could be justified under current WTO rules.[104] Doubts exist as to their possible coverage under the general exceptions in Art. XX GATT 1994, whose introductory clause or *chapeau* requires that such measures do not result in "arbitrary or unjustifiable discrimination between countries where the same conditions prevail", conditions that, in terms of carbon emissions reduction, may be very different depending on the principle of common but differentiated responsibilities.[105]

It is often considered that the most viable alternative to try to justify a border carbon tax is to design it under the concept of "border tax adjustments" referred to in Art. II:2(a) GATT 1994.[106] Under this provision, any WTO member may impose, "at any time on the importation of any product", "a charge equivalent to an internal tax imposed . . . in respect of the like domestic product or in respect of an article from which the imported product has been manufactured or produced in whole or in part".

A traditional example of a border tax adjustment relates to gold. For instance, if a country levies a specific tax on the production of gold, when imported gold or imported products containing gold (for instance, watches) arrive, they may be subject not only to the ordinary tariff at customs, but also to the corresponding tax on gold, which would be levied on domestic and imported gold alike in accordance with the principle of national treatment on internal taxation.

In order to design a carbon border tax that fits the notion of a border tax adjustment, the EU should provide for a levy that taxes one or more domestic products (e.g., steel or cement manufactured within the EU) at a rate based on the carbon emitted during their production process. This may be difficult, as the EU has so far preferred to reduce carbon emissions through obligations to limit GHG emissions rather than through taxes of this sort. In any case, a border tax adjustment always requires, by definition, a correspondence between the taxation imposed on domestic products and the tax imposed at the border on imported products.

In any event, the main legal obstacle to establish a carbon border tax in light of the WTO rules on border tax adjustments is that the definition of such adjustments refers to taxes charged on a "product" (such as gold) or an "article from which the imported product has been manufactured or produced in whole or in part", which has traditionally been understood as an ingredient physically present in the product (such as any gold contained in watches). In the case of carbon border taxes, what is taxed is not the product as such nor the ingredients physically incorporated within the product, but the carbon emissions that were generated during its production in the country of origin, which leave no physical trace in the final product (e.g., steel and cement). Pauwelyn argues that Art. II:2(a) GATT 1994 can be interpreted to cover taxes relating to the carbon footprint of a product as well, since, although carbon emissions are not physical ingredients of the product, they are also an item used in its production.[107]

Given this persistent legal uncertainty in the multilateral trading system, the EU can be assertive and carefully design a carbon border tax.[108] Although other countries will likely challenge such a measure through the WTO dispute settlement system, there are strong arguments to defend the compatibility of such taxes with WTO rules duly interpreted in light of the objective of sustainable development.

7. Final remarks

The EU is increasingly committed to using trade negotiations and measures to promote sustainable development and environmental protection on a global scale. In view of the various instruments and measures discussed in this chapter, some reflections and proposals can be put forward in this respect.

Regarding the GSP+, there is interest in continuing to collect data to better verify that beneficiary countries, in addition to accepting the relevant multilateral environmental agreements, are promoting their effective implementation. As the number of beneficiary countries of the GSP+ is currently small, it is worth considering the possibility of extending some aspects of environmental conditionality to the Standard GSP and even to the specific EBA arrangement granted to LDCs, thereby conveying the message that any tariff preference regime applied by the EU is linked to respect for certain basic environmental standards. In any case, this would require an increase in the EU's capacities to monitor a much higher number of countries and more instruments of dialogue and support.

As for RTAs, the systematic incorporation of a TSD chapter is welcome. It stresses that such preferential access to the EU market must be linked to compliance with certain international social and environmental standards. At the same time, it corrects the notable isolation that international trade agreements have traditionally had with respect to other sectors of international law, making explicit normative interconnections that promote a less fragmented and more systemic view of the international legal order. In any case, it is worth insisting on the monitoring of compliance with such chapters and pondering whether it is effective for these chapters to have a specific mechanism for the resolution of disputes that is not only different from the one provided for in other chapters of RTAs but also softer, since this conveys the idea that they have a different weight.

In the area of public procurement, the EU has made a decisive contribution to updating the WTO rules on the subject, currently contained in the GPA 2012 and probably facilitated by the fact that it is a plurilateral agreement. The GPA 2012 more clearly recognizes the possibility of subjecting public procurement to environmental requirements and explicitly takes up the notion of sustainable public procurement, promoting a work programme on the issue. The EU and its member states should demonstrate good practices for other countries that are now parties to the GPA 2012 or that might eventually join in the future.

In the area of anti-dumping measures, recent reforms in the EU's basic AD Regulation have incorporated social and environmental considerations for the first time. China, Russia, and other WTO members have criticized these reforms, viewing them as incompatible with the existing WTO AD Agreement. However, for the time being, these new EU rules have not been challenged through the WTO dispute settlement system and, if necessary, the EU could try to defend an evolving interpretation of the notion of dumping and of the criteria for the determination of normal value, which would take into account the "significant distortions" of certain markets, caused, among other factors, by low social and environmental standards. Ideally, the WTO AD Agreement could be updated to adapt its text to the new realities and challenges of the international community better, although the leadership of the EU as a global normative actor faces enormous difficulty in this particular area, where its approaches clash with those of other great powers.

As for trade measures concerning carbon footprint, the EU has already taken steps in this direction within the framework of its regulation on the classification of biofuels as sustainable or not, taking into consideration emissions in the production process and negative effects from ILUC. These measures are controversial. Some developing countries adversely affected by the measures have challenged them through the WTO dispute settlement system, but the adjudicative bodies have not yet ruled on their compatibility with the multilateral trading rules. The EU has also recently put forward a proposal to establish a carbon border tax as one of the flagship measures of "The European Green Deal". Although legal uncertainties remain as to whether all these measures relating to carbon footprint are compatible with WTO law, the EU can act in an assertive manner, carefully designing such measures in light of WTO jurisprudence on other types of trade measures relating to npr-PPMs and advocating an interpretation of multilateral trade agreements in line with the objective of sustainable development.

Overall, the EU's growing use of trade instruments and measures to promote sustainable development is welcome and should be continued. However, it seems clear that the purpose of such measures is not only the promotion of social and environmental standards on a global scale per se, but is also closely linked to a willingness to protect companies and jobs within the EU against tough international competition in order to ensure a more level playing field. Of course, it is understandable that the EU tries to promote fair trade not only to export its values, but also to protect its economic interests, without falling into the temptation to lower (or not raise) its own standards.

In any case, in order to strengthen the EU's leadership in an increasingly multipolar and complex international context, it is essential that all these EU trade tools are designed and implemented so as to comply with the provisions and procedures of the multilateral trading system. Indeed, the ideal would be to carry out a profound reform of WTO law and, following the model of the RTAs that the EU has recently been promoting, to incorporate many more social and environmental provisions, making it less focused on trade liberalization and more holistic in nature. However, while such a profound reform of the WTO seems urgent today, it also appears very difficult.

In this context, the EU must be imaginative, assertive, and respectful of current international rules and procedures, so as not to fall into aggressive unilateralism or be accused of practising new forms of disguised protectionism. The EU must therefore design its trade measures to promote sustainable development carefully, promoting dialogue with the countries and operators most affected and seeking to accompany its trade measures with other support measures, especially for LDCs and other vulnerable developing countries.

Notes

1 Krämer, 2013; Wouters et al., 2015; Douma, 2017; Hadjiyianni, 2019.
2 Art. 3.5 TEU.
3 Bhagwati, 1995; Bhagwati and Hudec, 1996; Raynolds et al., 2007.
4 Such as those contemplated, in particular, in Art. 21.2 TEU on sustainable development and trade (sections d, e and f).
5 Resolution 70/1 of the UN General Assembly, 25.9.2015, adopting the 2030 Agenda for Sustainable Development, with the SDGs.
6 EU Commission Communication: Trade for All – Towards a more responsible trade and investment policy, COM(2015)497 final, 14.10.2015.
7 Malmström, C.: "Foreword to the Trade for All Communication", October 2015, available from: https://trade.ec.europa.eu/doclib/docs/2015/october/tradoc_153846.pdf [accessed 18 March 2020].
8 COM(2015)497 final, *loc. cit.* note 6, p. 14, para. 4.1.1 and p. 15, para. 4.2.
9 *Ibid.*, p. 20, para. 5.1.
10 WTO: "Environmental Goods Agreement (EGA)", 2020, available from: www.wto.org/english/tratop_e/envir_e/ega_e.htm [accessed 18 March 2020].
11 Van den Bossche, 2020, pp. 1–2.
12 EU Commission: "EU concept paper on WTO reform", 2018, with a Section (I.C, p. 6) devoted to the possible contribution of the WTO to "the sustainability objectives of the global community", available from: https://trade.ec.europa.eu/doclib/docs/2018/september/tradoc_157331.pdf [accessed 18 March 2020].
13 EU Commission: "2018 EU Trade Defence Report", COM(2019)158 final, 27.3.2019, pp. 6 and 13.
14 Von der Leyen, U.: "Political Guidelines for the next European Commission 2019–2024", 2019, p. 17, available from: https://ec.europa.eu/commission/sites/beta-political/files/political-guidelines-next-commission_en.pdf [accessed 18 March 2020].

15 Von der Leyen, U.: "Mission letter to Phil Hogan, Commissioner-designate for Trade", Brussels, 20.9.2019, available from: https://ec.europa.eu/commission/sites/beta-poli tical/files/mission-letter-phil-hogan-2019_en.pdf [accessed 18 March 2020].

16 EU Commission Communication: The European Green Deal, COM(2019)640 final, 11.12.2019, pp. 20–21.

17 Soomin and Shirley, 2019.

18 WTO: "Enabling Clause – Differential and More Favourable Treatment Reciproc-ity and Fuller Participation of Developing Countries", Decision of 28.11.1979, Doc. L/4903.

19 OJ No. L 346, 31.12.2001.

20 WTO: *EC – Tariff Preferences*, Request for consultations by India, WT/DS246/1, 12.3.2002.

21 WTO: *EC – Tariff Preferences*, Appellate Body Report, WT/DS246/AB/R, 7.4.2004, para. 163.

22 *Ibid.*

23 OJ No. L 161, 30.6.2005.

24 OJ No. L 303, 12.10.2012.

25 EU Commission: "List of GSP beneficiary countries (as of 1 January 2019)", 2019, available from: https://trade.ec.europa.eu/doclib/docs/2019/may/tradoc_157889.pdf [accessed 18 March 2020].

26 *Ibid.*

27 Development Solutions: "Mid-Term Evaluation of the EU's Generalised Scheme of Preferences (GSP) Final Report", 2018, p. 9, available from: https://trade.ec.europa. eu/doclib/docs/2018/october/tradoc_157434.pdf [accessed 18 March 2020].

28 European Parliament: "Resolution of 14 March 2019 on the Implementation of the GSP Regulation (EU) No 978/2012 (2018/2107(INI))", available from: www.europarl. europa.eu/doceo/document/TA-8-2019-0207_EN.html [accessed 18 March 2020].

29 EU Commission: "2019 Report on Implementation of EU Free Trade Agreements (1 January 2018–31 December 2018)", 2019, available from: https://trade.ec.europa.eu/ doclib/docs/2019/october/tradoc_158387.pdf [accessed 18 March 2020].

30 EU-South Korea Free Trade Agreement, applied since 1 July 2011. *Ibid.*, p. 5.

31 EU-Colombia-Peru-Ecuador Trade Agreement, applied since 1 March 2013 for Peru; since 1 August 2013 for Colombia; since 1 January 2017 for Ecuador. *Ibid.*

32 EU-Central America Association Agreement, trade pillar applies, since 1 August 2013, to Honduras, Nicaragua and Panama; since 1 October 2013, to Costa Rica and El Sal-vador; since 1 December 2013, to Guatemala. *Ibid.*

33 EU-Canada Comprehensive Economic and Trade Agreement (CETA), provisionally applied since 21 September 2017. *Ibid.*

34 The EU-Singapore draft trade and investment agreements were signed on 19 Octo-ber 2018 and entered into force on 21 November 2019. See European Commission: "Overview of FTA and other trade negotiations – Updated February 2020", 2020, p. 2, available from: https://trade.ec.europa.eu/doclib/docs/2006/december/tradoc_118238. pdf [accessed 18 March 2020].

35 The EU-Japan Economic Partnership Agreement came into force on 1 February 2019. *Ibid.*

36 The EU-Vietnam trade and investment agreements were signed on 30 June 2019 and the European Parliament gave its consent to both agreements on 12 February 2020. *Ibid.*, p. 3.

37 The EU-Mercosur agreement in principle was reached on the trade part on 28 June 2019. Both sides are engaged in the legal review of the agreement. *Ibid.*, p. 4.

38 Von der Leyen, U.: "Political Guidelines . . .", *loc. cit.* note 14, p. 17.

39 EU Commission: "EU-Mercosur trade agreement: The Agreement in Principle and its texts", 12.7.2019, available from: https://trade.ec.europa.eu/doclib/press/index.cfm? id=2048 [accessed 18 March 2020].

40 Leal-Arcas and Wilmarth, 2015; Raess et al., 2018; Hradilová and Svoboda, 2018.
41 Ghiotto and Echaide, 2019.
42 Riaboi, 2020.
43 EU Commission: "Feedback and way forward on improving the implementation and enforcement of Trade and Sustainable Development chapters in EU Free Trade Agreements", Non-paper of the Commission Services, Brussels, 26.02.2018, available from: http://trade.ec.europa.eu/doclib/docs/2018/february/tradoc_156618.pdf [accessed 18 March 2020].
44 *Ibid.*, pp. 2, 3.
45 WTO: "WTO and Government Procurement", 2020, available from: www.wto.org/english/tratop_e/gproc_e/gproc_e.htm [accessed 18 March 2020].
46 *Ibid.*
47 WTO: "GPA – Parties, Observers and Accessions", available from: www.wto.org/english/tratop_e/gproc_e/memobs_e.htm [accessed 18 March 2020].
48 WTO: "Government Procurement Agreement – Opening Markets and Promoting Good Governance", 2015, p. 7, available from: www.wto.org/english/thewto_e/20y_e/gpa_brochure2015_e.pdf [accessed 18 March 2020].
49 Arrowsmith and Kunzlik, 2009; Caranta and Trybus, 2010; Arrowsmith and Anderson, 2011.
50 Art. X:6 GPA 2012.
51 WTO: "Annex E to Appendix 2 of the Decision on the Outcomes of the Negotiations Under Article XXIV:7 of the Agreement on Government Procurement", adopted on 30.3.2012, GPA/113, p. 444.
52 OJ No. L 94/65, 28.3.2014.
53 Sjåfjell and Wiesbrock, 2015.
54 WTO: "Symposium on Sustainable Procurement", Geneva, 22 February 2017, available from: www.wto.org/english/tratop_e/gproc_e/gp_symp_22feb17_e.htm [accessed 18 March 2020].
55 Baschuk, 2020.
56 OJ No. L 176, 30.6.2016.
57 OJ No. L 338, 19.12.2017.
58 OJ No. L 143, 7.6.2018.
59 The latest consolidated version of the EU's basic AD Regulation is available from: http://data.europa.eu/eli/reg/2016/1036/2018-06-08 [accessed 18 March 2020].
60 On the notions of social dumping and ecological dumping, see, for instance: Rauscher, 1994; Golub, 1997.
61 A legal basis for this special regime for NME countries is found in Art. 2.7 AD Agreement and the Second Supplementary Provision to Para. 1 of Art. VI in Annex I to GATT 1994. See Sohn, 2005.
62 Detlof and Fridh, 2007.
63 Para. 15 of the Protocol on the Accession of China to the WTO, WT/L/432, 23.11.2001, pp. 8–9.
64 On divergent interpretations of the scope of Para. 15 of the aforementioned Protocol on the Accession of China, see Battacharya, 2017.
65 Both cases, brought before the WTO by China on 12 December 2016, are designated, respectively, as *US – Price Comparison Methodologies* (WT/DS515) and *EU – Price Comparison Methodologies* (WT/DS516).
66 Art. 2(6a)(a) codified basic AD Regulation.
67 *Ibid.*
68 Art. 2(6a)(b) codified basic AD Regulation.
69 Art. 7(2d) codified basic AD Regulation.
70 EU Commission: Commission Staff Working Document, On Significant Distortions in the Economy of the People's Republic of China for the Purposes of Trade Defence Investigations, SWD(2017)483 final/2, 20.12.2017.

71 MOFCOM: "MOFCOM Spokesman Comments on the EU's Release of the Amend-ments to the New Anti-dumping Investigation Methodology, December 21, 2017", 2017, available from: http://english.mofcom.gov.cn/article/newsrelease/policyreleasing/201712/20171202691119.shtml [accessed 18 March 2020].

72 EU Commission: "The EU's New Trade Defence Rules and First Country Report", 20.12.2017, available from: https://ec.europa.eu/commission/presscorner/detail/en/MEMO_17_5377 [accessed 18 March 2020].

73 WTO Committee on Anti-Dumping Practices: "Minutes of the Regular Meeting Held on 25 April 2018", G/ADP/M/54, 3.8.2019, p. 3, para. 14.

74 *Ibid.*, pp. 3–4, paras. 12, 18 and 21.

75 *Ibid.*, p. 3, para. 19.

76 *Ibid.*, p. 4, para. 22.

77 Suse, 2017. See also, with a similar approach, Li, 2019.

78 Bonet, 2019, p. 166.

79 Marín Aís, 2019, p. 237.

80 Shadikhodjaev, 2018. Other publications questioning the compatibility of the EU's new basic AD Regulation with the WTO rules are Tietje and Sacher, 2018; Huyghe-baert, 2019.

81 According to Art. 3.2 of the WTO's Dispute Settlement Understanding (DSU), WTO agreements must be interpreted "in accordance with customary rules of interpretation of public international law". The evolving interpretation of WTO agreements has recently been analysed by Marceau, 2018.

82 Art. 31 VCLT.

83 EU Commission: "2018 EU Trade Defence Report", *loc. cit.* note 13, pp. 6 and 7.

84 Charnovitz, 2002; Conrad, 2011; Young, 2014; Cooreman, 2017; Maggio, 2017; Sifonios, 2018.

85 WTO: *US – Shrimp*, Appellate Body Report, WT/DS58/AB/R, 12.10.1998, and WTO: *US – Shrimp*, Art. 21.5 DSU Appellate Body Report, WT/DS58/AB/RW, 22.11.2001. See Yavitz, 2001–2002; Howse, 2002.

86 See chapters 5 and 8.

87 On the continuing legal uncertainties of certain trade measures to combat climate change in light of WTO law, see Condon, 2009; Cottier et al., 2009; Park, 2016; Delimatsis, 2016; Dobson, 2018.

88 OJ No. L 140, 5.6.2019.

89 WTO: *EU and certain Member States – Certain measures on the importation and marketing of biodiesel and measures supporting the biodiesel industry,* Request for consultations by Argentina, WT/DS459/1, 23.5.2013.

90 *Ibid.*, p. 3.

91 *Ibid.*

92 *Ibid.*

93 In any case, some authors have reflected on various legal aspects of this dispute, such as Grigorova, 2015.

94 WTO: *EU – Certain measures concerning palm oil and oil palm crop-based biofuels,* Request for consultations by Indonesia, WT/DS593/1, 16.12.2019.

95 European External Action Service: "Palm Oil Facts and Figures on Trade and Sustainability", Fact Sheet PO-01, 4.9.2019.

96 WTO, *EU – Certain measures concerning palm oil and oil palm crop-based biofuels,* Request for consultations by Indonesia, *loc. cit.* note 94, pp. 1–2.

97 OJ No. L 328, 21.12.2018.

98 OJ No. L 133, 21.5.2019.

99 WTO, *EU – Certain measures concerning palm oil and oil palm crop-based biofuels,* Request for consultations by Indonesia, *loc. cit.* note 94, p. 4, para. 4.

100 *Ibid.*, p. 6.

101 WTO: *EU – Certain measures concerning palm oil and oil palm crop-based bio-fuels*, Acceptance by the EU of the requests to join consultations, WT/DS593/8, 24.1.2020.
102 Von der Leyen, U.: "Political Guidelines . . .", *loc. cit.* note 14, p. 5; COM(2019)640 final, *loc. cit.* note 16, p. 5.
103 Mehling et al., 2019.
104 Sindico, 2008; Veel, 2009; Kaufmann and Weber, 2011; Pauwelyn, 2013; Holzer, 2014; Pirlot, 2017; Trachtman, 2017; Porterfield, 2019; Will, 2019.
105 Morosini, 2010; Davidson Ladly, 2012.
106 Pauwelyn, 2012, p. 51.
107 *Ibid.*, pp. 27–29.
108 As proposed, for example, by Krenek, 2020.

References

Arrowsmith, S. and Anderson, R.D. (Editors), 2011. *The WTO Regime on Government Procurement: Challenge and Reform*. Cambridge: WTO Publications and Cambridge University Press.

Arrowsmith, S. and Kunzlik, P. (Editors), 2009. *Social and Environmental Policies in EC Procurement Law – New Directives and New Directions*. Cambridge: Cambridge University Press.

Baschuk, B., 2020. How Government Contracts Became Next Trade War Front. *Bloomberg News*, 6 February 2020. Available from: www.bloomberg.com/news/arti cles/2020-02-06/how-government-contracts-became-next-trade-war-front-quicktake [Accessed 18 March 2020].

Battacharya, R., 2017. Three Viewpoints on China's Non-Market Economy Status. *Trade Law and Development*, 9(2), 188 ff.

Bhagwati, J.N., 1995. Trade Liberalisation and 'Fair Trade' Demands: Addressing the Environmental and Labour Standards Issues. *The World Economy*, 18(6), 745–759.

Bhagwati, J.N. and Hudec, R.E. (Editors), 1996. *Fair Trade and Harmonization: Prerequisites for Free Trade?* Cambridge, MA: The MIT Press.

Bonet, J., 2019. *La internormatividad entre las dimensiones económica y social del ordenamiento jurídico internacional: un espacio jurídico para la efectividad de los derechos económicos, sociales y culturales?* Barcelona: Huygens.

Caranta, R. and Trybus, M. (Editors), 2010. *The Law of Green and Social Procurement in Europe*. Copenhagen: DJØF Publishing.

Charnovitz, S., 2002. The Law of Environmental 'PPMs' in the WTO: Debunking the Myth of Illegality. *Yale Journal of International Law*, 27(1), 59–110.

Condon, B., 2009. Climate Change and Unresolved issues in WTO Law. *Journal of International Economic Law*, 12(4), 895–926.

Conrad, C.R., 2011. *Processes and Production Methods (PPMs) in WTO Law: Interfacing Trade and Social Goals*. Cambridge: Cambridge University Press.

Cooreman, B., 2017. *Global Environmental Protection through Trade: A Systematic Approach to Extraterritoriality*. Cheltenham: Edward Elgar Publishing.

Cottier, T., Nartova, O. and Bigdeli, S.Z. (Editors), 2009. *International Trade Regulation and the Mitigation of Climate Change*. Cambridge: Cambridge University Press.

Davidson Ladly, S., 2012. Border Carbon Adjustments, WTO-Law and the Principle of Common but Differentiated Responsibilities. *International Environmental Agreements*, 12, 63–84.

Delimatsis, P. (Editor), 2016. *Research Handbook on Climate Change and Trade Law*. Cheltenham: Edward Elgar Publishing.

Detlof, H. and Fridh, H., 2007. The EU Treatment of Non-Market Economy Countries in Anti-dumping Proceedings. *Global Trade and Customs Journal*, 2(7/8), 265–281.

Dobson, N.L., 2018. The EU's Conditioning of the 'Extraterritorial' Carbon Footprint: A Call for an Integrated Approach in Trade Law Discourse. *Review of European, Comparative and International Environmental Law*, 27(1), 75–89.

Douma, W.T., 2017. The Promotion of Sustainable Development through EU Trade Instruments. *European Business Law Review*, 28(2), 197–216.

Ghiotto, L. and Echaide, J., 2019. *Analysis of the Agreement between the European Union and the Mercosur*. Berlin: The Greens/EFA and Power Shift. Available from: www.annacavazzini.eu/wp-content/uploads/2020/01/Study-on-the-EU-Mercosur-agreement-09.01.2020-1.pdf [Accessed 18 March 2020].

Golub, S.S., 1997. Are International Labor Standards Needed to Prevent Social Dumping? *Finance and Development*, 34, 20–23.

Grigorova, J., 2015. EU's Renewable Energy Directive Saved by GATT Art. XX? Reflections on the Provisional Justification of Sustainability Criteria under GATT Art. XX in the Recent WTO Case 'European Union and Certain Member States – Certain Measures on the Importation and Marketing of Biodiesel and Measures Supporting the Biodiesel Industry' (DS459). *Oil, Gas and Energy Law Intelligence*, 13(3), 1–21.

Hadjiyianni, I., 2019. *The EU as a Global Regulator for Environmental Protection: A Legitimacy Perspective*. Oxford: Hart Publishing.

Hradilová, K. and Svoboda, O., 2018. Sustainable Development Chapters in the EU Free Trade Agreements: Searching for Effectiveness. *Journal of World Trade*, 52(6), 1019–1042.

Holzer, K., 2014. *Carbon-related Border Adjustment and WTO Law*. Cheltenham: Edward Elgar Publishing.

Howse, R., 2002. The Appellate Body Rulings in the Shrimp/Turtle Case: A New Legal Baseline for the Trade and Environment Debate. *Columbia Journal of Environmental Law*, 27, 491 ff.

Huyghebaert, K., 2019. Changing the Rules Mid-Game: The Compliance of the Amended EU Basic Anti-Dumping Regulation with WTO Law. *Journal of World Trade*, 53(3), 417–432.

Kaufmann, C. and Weber, R.H., 2011. Carbon-Related Border Tax Adjustment: Mitigating Climate Change or Restricting International Trade? *World Trade Review*, 10(4), 497–525.

Krämer, L., 2013. Exporting EU Environmental Product Standards to Third Countries. In Douma, W.T. and van der Velde, S. (Editors), *EU Environmental Norms and Third Countries: The EU as a Global Role Model?* The Hague: CLEER Working Papers 2013/5 – TMC Asser Institute, 19–33.

Krenek, A., 2020. How to Implement a WTO-Compatible Full Border Carbon Adjustment as an Important Part of the European Green Deal. *Österreichische Gesellschaft für Europapolitik (ÖGfE) Policy Brief*, 17 January 2020. Available from: https://oegfe.at/2020/01/wto-compatible-bca-green-deal/ [Accessed 18 March 2020].

Leal-Arcas, R. and Wilmarth, C.M., 2015. Chapter 5: Strengthening Sustainable Development through Regional Trade Agreements. In Wouters, J. et al. (Editors), *Global Governance through Trade – EU Policies and Approaches*. Cheltenham: Edward Elgar Publishing, 92–123.

Li, C., 2019. Relabeling Countries with Non-Market Economies as Countries with Market-Distorted Economies: The New EU Methodology in Anti-dumping Investigations at the Heart of a Storm. *Taiwanese Journal of Political Science*, 80, 37–62.

Maggio, A.R., 2017. *Environmental Policy, Non-Product Related Process and Production Methods and the Law of the World Trade Organization*. Cham: Springer.

Marceau, G., 2018. Evolutive Interpretation by the WTO Adjudicator. *Journal of International Economic Law*, 21(4), 791–813.

Marín Aís, J.R., 2019. Capítulo 8: La condicionalidad social en la Política Comercial Común. In Hinojosa Martínez, L.M. and Martín Rodríguez, P.J. (Editors), *International Markets Regulation and the Erosion of the European and Political Social Model*. Cizur Menor: Thomson Reuters Aranzadi, 207–238.

Mehling, M. et al., 2019. What a European 'Carbon Border Tax' Might Look Like. *Vox Column*, 10 December 2019. Available from: https://voxeu.org/article/what-european-carbon-border-tax-might-look [Accessed 18 March 2020].

Morosini, F., 2010. Trade and Climate Change: Unveiling the Principle of Common but Differentiated Responsibilities from the WTO Agreements. *The George Washington International Law Review*, 42, 713–748.

Park, D.Y. (Editor), 2016. *Legal Issues on Climate Change and International Trade Law*. Cham: Springer.

Pauwelyn, J., 2012. Carbon Leakage Measures and Border Tax Adjustments Under WTO Law. *Social Science Research Network*. Available from: https://ssrn.com/abstract=2026879 [Accessed 18 March 2020].

Pauwelyn, J., 2013. Chapter 15: Carbon Leakage Measures and Border Tax Adjustments Under WTO Law. In Van Calster, G. and Prévost, D. (Editors), *Research Handbook on Environment, Health and the WTO*. Cheltenham: Edward Elgar Publishing, 448–506.

Pirlot, A., 2017. *Environmental Border Tax Adjustments and International Trade – Fostering Environmental Protection*. Cheltenham: Edward Elgar Publishing.

Porterfield, M.C., 2019. Border Adjustments for Carbon Taxes, PPMs, and the WTO. *University of Pennsylvania Journal of International Law*, 41(4), 1–41.

Raess, D., Schmieg, E. and Voituriez, T., 2018. *The Future of Sustainable Development Chapters in EU Free Trade Agreements*. Brussels: EU Publications.

Rauscher, M., 1994. On Ecological Dumping. *Oxford Economic Papers*, 46, 822–840.

Raynolds, L.T., Murray, D.L. and Wilkinson, J. (Editors), 2007. *Fair Trade – The Challenges of Transforming Globalization*. London: Routledge.

Riaboi, J., 2020. El rechazo del lobby verde al acuerdo entre la UE y el Mercosur. *El Economista*, 3 February 2020. Available from: www.eleconomista.com.ar/2020-02-el-rechazo-del-lobby-verde-al-acuerdo-entre-la-ue-y-el-mercosur/ [Accessed 18 March 2020].

Shadikhodjaev, S., 2018. Non-Market Economies, Significant Market Distortions, and the 2017 EU Anti-Dumping Amendment. *Journal of International Economic Law*, 21(4), 885–905.

Sifonios, D., 2018. *Environmental Process and Production Methods (PPMs) in WTO Law*. Cham: Springer.

Sindico, F., 2008. The EU and Carbon Leakage: How to Reconcile Border Adjustments with the WTO? *European Energy and Environmental Law Review*, 17(6), 328–340.

Sjåfjell, B. and Wiesbrock, A. (Editors), 2015. *Sustainable Public Procurement under EU Law – New Perspectives on the State as Stakeholder*. Cambridge: Cambridge University Press.

Sohn, C., 2005. Treatment of Non market Economy Countries under the World Trade Organization Anti-dumping Regime. *Journal of World Trade*, 39(4), 763–786.

Soomin, L. and Shirley, S., 2019. A New Type of Imperialism: The Global North's Weapon of Mass intervention and Eco-Imperialism. *Konfrontasi: Jurnal Kultural, Ekonomi Dan Perubahan Sosial*, 8(1), 12–21. Available from: www.konfrontasi.net/index.php/konfrontasi2/article/view/11 [Accessed 18 March 2020].

Suse, A., 2017. Old Wine in a New Bottle: The EU's Response to the Expiry of Section 15(a)(ii) of China's WTO Protocol of Accession. *Journal of International Economic Law*, 20(4), 951–977.

Tietje, C. and Sacher, V., 2018. The New Anti-Dumping Methodology of the European Union: A Breach of the WTO Law? In Burgenberg, M. et al. (Editors), *The Future of Trade Defence Instruments: Global Policy Trends and Legal Challenges*. Cham: Springer, 89–105.

Trachtman, J.P., 2017. WTO Law Constraints on Border Tax Adjustment and Tax Credit Mechanisms to Reduce the Competitive Effects of Carbon Taxes. *National Tax Journal*, 70(2), 469–494.

Van den Bossche, P., 2020. Between Hope and Despair. *Society of International Economic Law (SIEL) Newsletter*, 44, 1–2.

Veel, P.E., 2009. Carbon Tariffs and the WTO: An Evaluation of Feasible Policies. *Journal of International Economic Law*, 12(3), 749–800.

Will, U., 2019. *Climate Border Adjustments and WTO Law – Extending the EU Emissions Trading System to Imported Goods and Services*. Leiden: Brill and Nijhoff.

Wouters, J. et al. (Editors), 2015. *Global Governance through Trade – EU Policies and Approaches*. Cheltenham: Edward Elgar Publishing.

Yavitz, L., 2001–2002. The WTO and the Environment: The Shrimp Case That Created a New World Order. *Journal of Natural Resources and Environmental Law*, 16, 203 ff.

Young, M.A., 2014. Trade Measures to Address Environmental Concerns in Faraway Places: Jurisdictional Issues. *Review of European, Comparative and International Environmental Law*, 23(3), 302–317.

5 The EU's global leadership in the fight against illegal, unreported, and unregulated fishing

Xavier Pons Rafols

1. Introductory remarks

Illegal, unreported, and unregulated (IUU) fishing is an international phenomenon with serious environmental, economic, and social consequences. From the perspective of European Union's (EU) action in the global protection of the environment, this chapter sets out to analyse the fight against IUU fishing and the international impact of EU action. To a large extent, this effect derives from regulations adopted by the EU with regard to IUU fishing – in force since January 2010 – which, in addition to their effects at EU level, have had effects that are also international in scope, improving the compliance of third states with fisheries conservation and sustainable management measures. This important international influence has placed the EU in a leading role in the application of international measures for the conservation and management of living marine resources,[1] influencing in the adoption of the Agreement on Port State Measures (PSMA) within the framework of the United Nations Food and Agriculture Organization (FAO). Furthermore, the EU has taken an active role in new developments in a variety of other international institutions, reinforcing multilateral international action against IUU fishing.

To this end, the chapter turns next to a presentation of the global phenomenon of IUU fishing, followed by an analysis of its international legal conceptualization, the basis and scope of pertinent EU regulations, and the essential content of these regulations. On this basis, the international impact of these regulations is then analysed, particularly with respect to the mechanism of a list of non-cooperating third countries and the international effects of the EU's multilateral action on other international legal and institutional frameworks.

2. The global phenomenon of IUU fishing

IUU fishing is now a phenomenon of major international importance. The European Commission estimates that between 11 and 26 million tonnes of fish, i.e., just over 15% of the world's catches,[2] worth a minimum value of 10 billion euros per year, come from IUU fishing. This is a phenomenon that seriously threatens marine ecosystems and the global environment, with serious consequences for the conservation of ocean resources, and it undermines sustainable fisheries and food security, since it can lead to the total collapse of a fishery and seriously undermine efforts to recover depleted fish stocks.[3] It also has undoubted economic consequences, as it constitutes unfair competition for legal fishing activity and has negative effects on tax revenues.

In addition, IUU fishing affects developing countries and small-scale fishing communities, and it is associated with safety problems on board vessels and with forced labour practices that show no respect for existing labour standards in the fisheries sector. In many cases, it is also linked to transnational organized crime activities, such as trafficking in drugs, arms, or human beings, or it directly forms part of transnational organized fisheries crime.[4] Combating this type of fishing and its harmful effects on biodiversity and the social and economic sustainability of fisheries remains a fundamental part of fisheries governance, because it is a threat to the conservation of resources and their sustainability, the livelihoods of fishers and other fisheries actors, and it exacerbates malnutrition, poverty, and food insecurity. Hence, there is also a need for multilateral and multifaceted international action to address the phenomenon and all its connections.

The serious consequences of IUU fishing have meant that, among many other aspects of environmental, economic, and social sustainability, the Sustainable Development Goals (SDGs) adopted by the United Nations (UN) in 2015 underline the importance of taking action against IUU fishing to promote the conservation and sustainable use of oceans and marine ecosystems. More specifically, Target 14.4 of the SDGs aims to effectively regulate harvesting and end overfishing, IUU fishing, and destructive fishing practices and implement science-based management plans in order to restore fish stocks in the shortest time feasible, at least to levels that can produce maximum sustainable yield, as determined by their biological characteristics, by 2020.[5]

In this regard, it should be noted that there are rules to regulate fishing both in marine areas under state jurisdiction and in areas of the high seas, and that states derive obligations from these rules in relation to the control and regulation of fishing activities. However, insufficient control and enforcement by flag states, weak regulatory measures, subsidies that are harmful in producing fleet overcapacity in the fishing sector, and insufficient port and coastal state control have resulted in intensive and unregulated or directly illegal fishing activities. This is how the phenomenon of IUU fishing has grown in recent decades, particularly in weak or failed states,[6] and given rise to international actions to combat fishing of this kind, thus expressing the need for efforts by the entire international community.[7]

The effects of IUU fishing and the insufficiency of international instruments[8] led to the first international actions in the late 1990s, which resulted in the International Plan of Action to Prevent, Deter and Eliminate Illegal, Unreported and Unregulated Fishing (IPOA-IUU), adopted by FAO in 2001.[9] The IPOA-IUU, which is voluntary in scope, provides all states with comprehensive, effective, and transparent measures to combat IUU fishing, including regional fisheries management organizations or arrangements (RFMOs or RFMAs) and relevant regional fisheries bodies (RFBs), established under international law.[10] To this end, the

IPOA-IUU sets out strategies and measures directed at either all states, flag states, coastal states, or port states, together with conservation and management measures (CMMs) for RFMOs.[11]

In accordance with this international approach, the responsibilities in the fight against IUU fishing are, firstly, those of flag states, which are responsible for controls and authorizations on the performance of vessels flying their flag; secondly, those of port states, which are responsible for controlling the admission, entry, and unloading in ports; thirdly, those of coastal states, which are responsible for controlling fishing activities carried out in the waters under their jurisdiction; and lastly, those of trading states, which are responsible for ensuring that any fish entering their markets come from responsible fishing.[12]

From this initial perspective, it should be noted that the EU acts in each of these dimensions because it has exclusive competence for the conservation of marine biological resources under the Common Fisheries Policy (CFP)[13] and exclusive competence for trade policy. In addition, the EU is a member of several RFMOs and RFBs (six tuna organizations and 11 zonal organizations); it is a party to the United Nations Convention on the Law of the Sea (UNCLOS) and other international fisheries agreements; and it has adopted many bilateral and multilateral fisheries agreements.[14] The fight against IUU fishing from the EU perspective, however, responds not only to its competences, but also to its interests insofar as it has a wide seafront and its fishing fleet – especially in a number of countries, such as Spain – has an important structural dimension, which is powerful in terms of deep-sea fishing capacity, making it the third-largest extractive power in the world. Finally, the EU is also the world's largest importer of fishery products, hence its interest in fish supply chain traceability.[15]

3. International legal concept of IUU fishing

The IPOA-IUU also established the legal definition of the phenomenon to be combated. Despite certain ambiguities, this definition is still the most complete, current international definition of IUU fishing.[16] For reasons of space and in order not to literally reproduce the definitions of illegal fishing, unreported fishing, and unregulated fishing, suffice it to say that illegal fishing means, in essence, any fishing which is carried out – whether by national or foreign vessels, whether in waters under state jurisdiction or not – in violation of state regulations or of the international rules established by RFMOs to regulate this fishing for these vessels or in these spaces. In other words, illegal fishing is a fishing activity in violation of existing rules, whether these rules are national or international in nature. Unreported fishing is fishing that has not been reported or that has been reported inaccurately in violation of state or RFMO regulations. In other words, unreported fishing is simply a form of illegal fishing, since it is defined by the contravention of national or RFMO procedures in relation to the declaration of the activity undertaken or the catches made.

The concept of unregulated fishing is much more complex and confusing.[17] As defined by the IPOA-IUU, it comprises, in part, any activities that are carried out in an RFMO's area of application by vessels without nationality, or by vessels flying the flag of a state not party to that organization, or by a fishing entity in a manner that is not consistent with, or contravenes, the CMMs of that organization. The problem lies precisely in the fact that these vessels either have no nationality or fly the flag of a state that is not bound by these measures, and it is therefore difficult to clearly establish the illegality of their behaviour,[18] unless it is understood that there is an obligation to comply with these measures on the basis of the UNCLOS and the general duty of cooperation established by other international agreements, or that this compliance derives from a rule of customary international law, in which case we would be dealing with illegal fishing rather than unregulated fishing.[19]

Moreover, activities carried out in areas or for fish stocks in relation to which there are no applicable CMMs, and where such fishing activities are conducted in a manner inconsistent with state responsibilities for the conservation of living marine resources under international law, also constitute unregulated fishing.[20] However, while there is no specific regulation and the obligations under international law are too vague in relation to fishing in an unregulated area or on an unregulated stock, the International Tribunal for the Law of the Sea (ITLOS) has repeatedly found that states do have a clear duty to protect and preserve the marine environment, and that the conservation of the living resources of the sea is an element in the protection and preservation of the marine environment.[21]

In this sense, given the manner in which unregulated fishing is defined in the IPOA-IUU, we are faced with a concept that is elusive, since the fact that there is no corresponding regulation does not necessarily imply that the fishing activity is harmful, should be prosecuted, or should be considered illegal.[22] All this creates a confusing ambiguity, since it may well turn out that some IUU fishing, specifically some so-called unregulated fishing, does not constitute a violation of the applicable regional and international CMMs.

In short, we are faced with only two types of activities – illegal and unregulated fishing – that are lumped together, although they should receive differential treatment.[23] In any case, despite these ambiguities, the concept of IUU fishing aims to cover all forms of fishing that contribute to the impoverishment of fisheries resources, whether they take place outside international and national mechanisms aimed at ensuring that fishing is developed in a responsible manner or in contravention of such mechanisms.[24]

4. Basis and scope of EU regulations on IUU fishing

Initially, the EU's interests and competences led it to express a willingness to take the lead in the international fight against IUU fishing with the adoption by the European Commission, shortly after the adoption of the IPOA-IUU, of a Community Action Plan to eradicate IUU fishing,[25] which was then complemented by the development, implementation, and control of the CFP.[26] This is why, in addition to specific EU legislation on IUU fishing, reference should also be made to the EU fisheries control system to ensure compliance with the CFP and the EU's system for authorizing the fishing activities of EU vessels and the access of third-country vessels to EU waters, as well as EU measures aimed at conserving fish stocks in relation to countries that allow non-sustainable fishing.[27] All of these efforts should be viewed within the framework of the reform of the policy[28] and the institutional development undertaken through the creation of the European Fisheries Control Agency (EFCA).[29]

Against the background of the Action Plan, the Commission launched a new strategy which, in line with the EU's international commitments and aimed at strengthening the fight against IUU fishing, sought to have a direct impact on reducing the profit made from the trade in IUU fishing products, i.e., to act through a market approach.[30] This strategy was mainly embodied in two regulations, Regulation 1005/2008[31] and Commission Regulation 1010/2009,[32] which aimed to prevent, deter, and eliminate IUU fishing by ensuring that all states, both member states and third countries, complied with their obligations.[33] This comprehensive legal framework on IUU fishing, which is complementary to the other EU fisheries rules, took effect on 1 January 2010 and remains fully in force, with various implementing and updating rules adopted since then.

As for the material scope of these Regulations, it should be noted that they apply to IUU fishing, understood in the sense of the IPOA-IUU definition assumed by the EU, and supplemented with general criteria for the presumption of IUU fishing.[34] In any case, it should also be noted, on one hand, that the EU Regulations do not formulate the IPOA-IUU provision stipulating that certain unregulated fishing may take place in a manner that is not in violation of applicable international law, thus making the EU's conceptualization more restrictive. On the other hand, the Regulations do not solely refer to fishing activities in the strict sense, but also cover other activities that are associated with fishing, such as shipping, transhipment, processing, landing, trade, and distribution of fish and fishery products that may be derived from IUU fishing.[35] Thus, these other activities associated with IUU fishing are also likely to be considered as such for the purposes of inclusion on the list of IUU vessels and consequently be subject to the corresponding sanction measures.

In relation to their spatial scope of application, the European Regulations cover any fishing activity at sea, whether it is carried out in EU waters, i.e., waters under the jurisdiction of a member state, or in waters under the jurisdiction of a third state or in areas of the high seas, whether or not the latter areas are subject to a regulatory regime through an RFMO.[36] Consequently, they express a broad regulatory willingness to combat IUU fishing, which goes beyond the waters under the jurisdiction of member states and the CFP itself, thus having extraterritorial effects.[37]

Finally, following the same logic in relation to their subjective scope of application, the European Regulations apply not only to all fishing vessels flying the flag of a member state, but also to all fishing vessels flying the flag of third states. This subjective extension responds to a need for the transparent and non-discriminatory application of the measures, which may affect both vessels from third states and owners and operators from third states, as well as vessels and owners and operators from member states. From this perspective, the only requirement for the application of the Regulations is the existence of a connecting factor with the EU, either through access to ports of member states or the existence of trade flows of fishery products to or from the EU, or because the flag of the fishing vessels or the nationality of their owners and operators is that of a member state.

5. Core content of the EU Regulations on IUU fishing

European Regulations have a catch certification scheme that is the central operational instrument under the market approach and serves to ensure the traceability of fishery products. The requirement is that all trade in fishery products with the EU, whether imported or exported, must be accompanied by a catch certificate that has been validated by the flag state of the vessel carrying out the fishing activity, and that will serve to certify the legality of its catches.[38] In this respect, the burden of proof that fishery products marketed in the EU are not derived from IUU fishing lies with the fishing vessel and the flag state. Through this kind of added extraterritorial effect from European Regulations, the implication is that any third-party flag state of a fishing vessel must have its own national enforcement and control regime so that its catch certificates can be validated and accepted by the EU. The EU does not establish new criteria or CMMs for fisheries additional to those already existing at international level; it only requires that, for entry by third-state fishing vessels into the European market for fishery products, these third states must have adequate verification and control mechanisms in place to ensure compliance with their internal rules and existing international standards.

Another important development is the establishment of an enhanced system of control and inspection of third-country fishing vessels by port state authorities. This is in line with the new international approach to port state measures and took shape in 2009 with the PSMA. The issue is that EU Regulations have generally established a comprehensive system for monitoring the legality of catches by EU fishing vessels,[39] whereas the system applicable to fishery products caught by third-state fishing vessels and imported into EU territory has not allowed for a similar level of control. The Regulations on IUU fishing now reinforce the rules governing access to EU ports by third-state fishing vessels to ensure adequate control of the legality of any products landed by these vessels, too.

In this respect, fishing vessels of third states must fulfil several conditions to gain permission to enter the designated ports of member states, receive port services, and carry out landing or transhipment operations in said ports. Specifically, the conditions require prior notification of entry to and use of a port, accompanied, where appropriate, by the relevant validated catch certificate, while they prohibit all transhipment operations between third-state fishing vessels, and between third-state fishing vessels and fishing vessels flying the flag of a member state in EU waters, requiring instead that any such transhipments may take place only in port. If the information is complete – and, in the case of fishery products, accompanied by the relevant catch certificate – then entry into designated ports may be authorized to third-state fishing vessels.

In addition to prior notification, EU Regulations also provide for a strict procedure for port inspections and even set a minimum threshold for such inspections. Inspections are based on risk management and, to this end, Regulation 1010/2009 has established common risk management criteria as a benchmark for port inspections and verifications of catch certificates.[40] In any case, vessels for which there are reasonable grounds to suspect that they may be involved in IUU fishing activities, *inter alia* because they have been sighted or are on the list of suspected vessels of an RFMO, should be systematically inspected.[41] If inspection reveals evidence of IUU fishing, the Regulation stipulates that the competent authorities of the port member state shall not authorize the landing or transhipment of catches and shall make various notifications and requests for cooperation, depending also on the place where the suspected infringement has been committed.

Finally, and also in relation to what has just been indicated, EU Regulations establish a warning system and mutual assistance between member states to advise of the presence of IUU fishing, and they set out various coercive and sanctioning mechanisms, including the imposition of effective and dissuasive criminal and/or administrative sanctions.[42] Most relevant is the elaboration by the EU, on one hand, of a list of non-cooperating third countries – see next section – and, on the other hand, a list of vessels engaged in IUU fishing, on the basis of information obtained by the European Commission, by member states, or by a body designated by the European Commission.[43] The list of vessels will also include any vessels that have already been identified as IUU fishing and put on the lists adopted by the RFMOs concerned, with the consequence that the measures provided for in Regulation 1005/2008 will apply. In this regard, the identification by RFMOs of vessels engaged in IUU fishing will be immediately recognized and incorporated into the EU list of vessels engaged in IUU fishing.[44]

European legislation also provides for a wide range of measures applicable to fishing vessels on the list, including a ban on the import, export, and re-export of fishery products from IUU fishing vessels; a ban on access to EU ports for such vessels; the prohibition of cooperation, assistance, transhipment, and joint fishing operations between EU fishing vessels and vessels on the list; the prohibition of authorization to fish in EU waters for vessels flying the flag of a third state on the list; and the withdrawal of fishing authorizations or special fishing permits issued by flag member states.[45]

6. The international effect of EU action: establishment of a list of non-cooperating third countries

One of the aspects which most clearly demonstrates the scope and international impact of European Regulations on IUU fishing is precisely the establishment of a list of non-cooperating third states whereby, if appropriate, and following the Commission's notification to any third state under consideration as non-cooperating,[46] the third state in question may be registered by means of a Decision of the EU Council.[47] The system of listing non-cooperating third states constitutes a major qualitative leap forward, mainly because the identification of such third states is not directly linked to vessels flying their flag while carrying out IUU fishing activities exclusively, or to the export of fishery products derived from IUU fishing from a given state to the EU. Indeed, it should be noted that article 31.3 of Regulation 1005/2008 provides for a much more general formulation, whereby any third state which "fails to discharge the duties incumbent upon it under international law as flag, port, coastal or market state, to take action to prevent, deter and eliminate IUU fishing" may be considered a non-cooperating third state.

The formulation, therefore, is an open approach – on the basis of compliance with international law obligations – to the identification as such of a non-cooperating third state, including a broad interpretation of port state obligations.[48] Furthermore, among the elements that the European Commission will take into account in order to proceed with such an identification are, on one hand, "the ratification of, or accession of the third countries concerned to, international fisheries instruments", as well as, on the other hand, "the status of the third country concerned as a contracting party to regional fisheries management organizations, or its agreement to apply the conservation and management measures adopted by them".[49]

Under these rules, the European Commission has established a pre-identification procedure, also known as a "yellow card", by which it warns a state that it is at risk of being declared a non-cooperating state. From that point onwards, the EU provides assistance and collaboration to the pre-identified state so that it can remedy any shortcomings, and the EU establishes deadlines to remedy such shortcomings, which can be extended if necessary and if progress is made in adopting the appropriate measures and control mechanisms. In this regard, and by decision of the European Commission, any pre-identification can be lifted by means of a so-called "green card" if appropriate improvements have been made, thus ruling out the possibility of the state being placed on the list of non-cooperating states. If there are no improvements or no adequate cooperation, a third state which has been pre-identified by the European Commission as a non-cooperating state may be included, by decision of the Council, on the list of non-cooperating third states, by means of the identification procedure known as a "red card", which implies that sanctions, especially measures of a commercial nature, may be imposed.[50]

The inclusion of non-cooperating third states on the list is ultimately a unilateral claim by the EU to require compliance by third states with existing international rules and thus to assume powers of control over third states and the application of coercive measures of a commercial nature, which may be inequitable or focused only on certain states.[51] These measures are compatible with international law to the extent that the relevant international provisions also provide for mechanisms to ensure compliance, but in the absence of such mechanisms they may be controversial.[52] The fact is that these measures, firstly, derive directly from the fight against IUU fishing and, therefore, are justified by the non-adoption of measures by a third state to prevent, deter, and eliminate IUU fishing; secondly, they are linked to existing international instruments on the conservation and management of fisheries resources; thirdly, they are not understood as a trade barrier incompatible with international trade rules and can be covered by article XX of GATT

1994;[53] and, lastly, they clearly show the willingness of the EU to take the lead in this field.[54]

7. The multilateral influence of EU action: the port state measures agreement and other international legal and institutional frameworks to combat IUU fishing

The multilateral action and influence of the EU in relation to the fight against IUU fishing is carried out through a variety of international institutional frameworks, especially through the RFMOs and RFBs in which the EU participates as a full member and through certain specialized agencies of the UN. In this regard, it should be noted that in recent years there has been an increase in the number of Catch Documentation Schemes (CDSs) established in various RFMOs, with a scope similar to that of the EU and reinforced by the Voluntary Guidelines for CDSs adopted by the FAO Conference.[55] In addition, Voluntary Guidelines For Flag State Performance have been adopted within the framework of the FAO,[56] and since 2017 progress has been made on the Global Record of Fishing Vessels, Refrigerated Transport Vessels, and Supply Vessels, which will extend a Unique Vessel Identifier (UVI) as a global unique number that is assigned to a vessel to ensure traceability through the reliable, verified, and permanent identification of the vessel.[57] All these actions reinforce the multilateral fight against IUU fishing.

As noted previously, the multilateral action of the EU is carried out through three specialized agencies of the UN: FAO, International Maritime Organization (IMO), and International Labour Organization (ILO).[58] In this respect, there are three binding legal instruments: the already mentioned PSMA in a particularly significant way; the Cape Town Agreement of 2012 on the Implementation of the Provisions of the 1993 Protocol relating to the Torremolinos International Convention for the Safety of Fishing Vessels 1977; and the Work in Fishing Convention 2007. The three legal instruments and their thematic areas highlight the need for a joint, multilateral, and multifaceted approach in the fight against IUU fishing.

As regards the PSMA, it should be noted that the various initiatives promoted by the FAO in relation to measures for port states to take to combat IUU fishing[59] gradually evolved towards the adoption of a legally binding instrument. As part of this process, the EU participated actively in the Technical Consultation and as a full member of both the Committee on Fisheries and the FAO Conference. The PSMA was adopted on 22 November 2009[60] and entered into force on 5 June 2016, with 63 states parties in addition to the EU.[61] With respect to the substantive content of the PSMA,[62] it should be first stressed that it focuses exclusively on measures that can be taken by port states, whereas only in a more indirect and tangential manner does it address issues related to the rights and obligations of other states. The PSMA, therefore, is conceived as only one pillar in a broader strategy to combat IUU fishing that must continue to evolve, and it addresses the behaviour of flag states as one of its central axes. This contrasts with the much broader approach of EU Regulations, which aim to cover all dimensions of the IUU phenomenon by establishing various types of obligations for member states.[63]

As to the PSMA's essential core, it should be stressed that it is made up of provisions relating to entry into port, the use of ports, and inspections and follow-up actions. In addition to the designation of ports by the parties to the PSMA, it establishes requirements such as the prior application for entry into port with any information required by the port state to authorize a vessel's entry. It remains up to each port state, by virtue of its sovereign powers, to decide whether or not to authorize entry into port. In relation to inspections and follow-up actions, the PSMA provides that each party shall inspect the number of vessels required to achieve an annual level of inspections in its ports sufficient to achieve the objectives of the Agreement. The PSMA also establishes the priorities for inspection. In short, it can be said that the PSMA is certainly limited in scope and is a minimum agreement, but nothing prevents the adoption of other measures by a port state, such as the stricter and more precise measures with specific sanctioning procedures provided for in European Regulations or any other measures that the flag state of a vessel has expressly requested or has consented to, provided that they are in accordance with international law.

This is also the approach that the EU supported during the process of elaboration of the PSMA and that was clearly expressed in the Technical Consultation,[64] thus giving cover to the more complete and strict content of EU Regulations on IUU fishing. It is precisely the more complete nature of the EU Regulations that can also exert influence in the implementation of the PSMA, just as the active participation of the EU had an influence in its very adoption.[65] In this sense, there is a certain interplay between the EU Regulations on IUU fishing and the PSMA, which will make it possible to verify the EU's capacity to influence, politically and legally, the progressive practical implementation of the provisions of the PSMA. In fact, it should be noted that the PSMA has not yet incorporated the establishment of general criteria for the presumption and determination of the existence of IUU fishing, as the EU has done. Nor have risk management criteria been established for the specific adoption of control and inspection measures in relation to IUU fishing, as the EU has done in linking such control and inspection actions to the precautionary approach and ecosystem-based approaches grounded on international CMMs. In this sense, the greater precision of European provisions may also have an emulation effect in relation to control and inspection measures, which could eventually lead to their widespread adoption.

As regards the Cape Town Agreement of 2012 on the Implementation of the Provisions of the 1993 Protocol relating to the Torremolinos International Convention for the Safety of Fishing Vessels 1977, it should be noted that the Agreement aims to ensure the safety of fishing vessels within the framework of the IMO's competences and activities. Since neither the 1977 Convention nor the 1993 Torremolinos Protocol had entered into force, the IMO encouraged the adoption of the Cape Town Agreement 2012, which partially relaxed the conditions for their entry into force.[66] The EU cannot become a party to these texts, which do not contain a regional economic integration organization clause. However, the material contents of both the 1977 Protocol and the 2012 Agreement do fall within the exclusive competence of the EU, as the Protocol was incorporated into EU law by

Directive 97/70/EC.[67] In view of this situation and considering the desirability of the entry into force of the Agreement to improve the safety of fishing vessels, a Council Decision was adopted in 2014 authorizing member states to sign, ratify, or accede to the Cape Town Agreement.[68] The Decision annexed the Declaration to be deposited by member states concerning the competence and legislation of the EU and called on member states to become parties to the Agreement within two years.

However, this process has not made substantial progress either at the European or international level,[69] and the IMO, in collaboration with the Spanish government, convened a Ministerial Conference on Fishing Vessel Safety and Illegal, Unreported and Unregulated (IUU) Fishing in 2019, thus highlighting the connection between vessel safety and the fight against IUU fishing.[70] The Declaration adopted at the Conference[71] aims to promote a favourable political process and to bring about the entry into force of the Cape Town Agreement by 2022. We will have to wait and see, but from the EU perspective, this would certainly improve the safety of people working on fishing vessels, modernize the technical requirements of these vessels, and contribute to the fight against IUU fishing.

Finally, mention should be made of ILO Convention No. 188, Work in Fishing, adopted in 2007 but not entering into force until ten years later, on 16 November 2017.[72] The difficulties of bringing the Work in Fishing Convention into force can be seen from the fact that, even today, only three of the 27 EU member states are bound by it, although it will soon enter into force for another four.[73] However, the crucial importance of the Convention in relation to the living and working conditions of workers in the fisheries sector, one of the most arduous and dangerous occupations, did lead to the adoption of Directive (EU) 2017/159.[74] The Directive is particularly noteworthy because it transforms into EU law – requiring the corresponding transposition by member states – an agreement between the social partners concerned in accordance with article 155.2 of the Treaty on the Functioning of the European Union (TFEU).

Specifically, the agreement between employers' and workers' organizations largely transposes the provisions of the ILO Convention into EU law, stipulating that member states must adopt the laws, regulations, and administrative provisions necessary for compliance by 15 November 2019. The process is also extraordinarily complex and meets with state resistance, but insofar as the EU cannot be a party to the Convention, this Directive – which arises from the social partners themselves – will serve to achieve the objectives of improving the living and working conditions and protecting the health and safety of workers in the fisheries sector, which will undoubtedly contribute to the fight against IUU fishing, too.

8. Final remarks

By way of final considerations, it should be noted, firstly, that significant progress has been made at the global level in the fight against IUU fishing since the adoption of the IPOA-IUU almost 20 years ago, and that EU action represents an important qualitative element of this progress. More than ten years after the entry into force of the European Regulations on IUU fishing, the general balance that can be made of these regulations is clearly positive. They have had an impact on the regulation of fishing activities around the world. At the internal level of the EU, there has also been an improvement in member states' fulfilment of their responsibilities as flag states, coastal states, port states, and trading states within the framework of CFP reform.

At the international level, EU action has led to the strengthening of port state obligations as a mechanism to combat IUU fishing. It has also prompted the establishment of a general methodology with a clear market approach that fundamentally incorporates the CDSs, so that all states end up complying with their international obligations in the fight against IUU fishing. From the EU perspective, a three-pronged approach has been adopted: all states must have enforcement and control bodies that are able to validate catch certificates if they intend for fishery products caught by fishing vessels flying their flag to be exported to the EU; the EU has committed itself to work intensively with third states to improve their compliance and strengthen their fisheries governance systems; and, finally, a list of non-cooperating third states has been established to register any third states that do not cooperate or do not properly control the fishing activities carried out under their jurisdiction.

In addition to the international effects of EU action, the EU's multilateral influence and action must be emphasized both in relation to the PSMA in particular – taking into account the EU's status as a regional economic integration organization belonging to the FAO – and in relation to the Cape Town Agreement for the Safety of Fishing Vessels and the ILO Convention on Work in Fishing. This is the way forward, but there is no doubt that more progress must and can be made. In this sense, finally, in addition to highlighting the EU's influence and international action from a multilateral perspective, it is clear that any complex phenomenon that is global in scope, such as IUU fishing, needs to be tackled at the international level through a joint and multifaceted approach that addresses the various dimensions of the phenomenon and contributes effectively, as in the present case, to prevent, deter, and eliminate IUU fishing in order to safeguard oceans and marine ecosystems.

Notes

1 Pons, 2016.
2 EU Commission Communication: On the application of Council Regulation (EC) No 1005/2008 establishing a Community system to prevent, deter and eliminate illegal, unreported and unregulated fishing, COM(2015)480 final, 1.10.2015, p. 1.
3 FAO, 2018, p. 110.
4 Stolsvik, 2019.
5 Transforming our world: the 2030 Agenda for Sustainable Development, Resolution 70/1, adopted by the General Assembly on 25 September 2015.
6 Rodrigo, 2019, p. 163.
7 Riddle, 2006, p. 266.
8 Basically, the Geneva Convention on fishing and conservation of the living resources of the high seas of 29 April 1958, UNTS, vol. 559, p. 285; the United Nations

Convention on the Law of the Sea of 10 December 1982, OJ No. L 179, 23.6.1998; the Agreement to Promote Compliance with International Conservation and Management Measures by Fishing Vessels on the High Seas, 24 November 1993, OJ No. L 177, 16.7.1996; and the Agreement for the Implementation of the Provisions of the United Nations Convention on the Law of the Sea of 10 December 1982 relating to the Conservation and Management of Straddling Fish Stocks and Highly Migratory Fish Stocks of 4 August 1995, OJ No. L 189, 3.7.1998. The EU is a contracting party to these last three agreements.

 9 Adopted by FAO Council on 23 June 2001 (CL 120/REP).

10 Van der Marel, 2019, p. 298.

11 Edeson, 2001.

12 The principle of responsible fisheries was proclaimed in 1995 in the FAO Code of Conduct for Responsible Fisheries (FAO Conference Resolution 4/95 of 31 October 1995, C 1995/REP).

13 Regarding IUU fishing, the Court of Justice of the European Union reiterated this point in its judgment in Case C-73/14, Council v. Commission, 6.10.2015. ECLI:EU:C:2015:663.

14 This is why the fight against IUU fishing worldwide is a priority for the EU, as stated in the Joint Communication of the European Commission and of the High Representative of the Union for Foreign Affairs on International Ocean Governance: an agenda for the future of our oceans, JOIN(2016) 49 final, 10.11.2016.

15 Pons, 2016, p. 186.

16 Paragraphs 3.1, 3.2 and 3.3 of the IPOA-IUU.

17 Theilen, 2013.

18 Serdy, 2017.

19 Van der Marel, 2019, pp. 295–296.

20 The dual condition required is that the fishing must take place in areas where there are no applicable conservation and management measures "and" that fishing activities must be carried out in a manner inconsistent with the responsibilities of states for the conservation of living marine resources.

21 For example, in Southern Bluefin Tuna (*New Zealand* v *Japan*; *Australia* v *Japan*), Provisional Measures, Order of 27 August 1999, ITLOS Reports 1999, p. 280, para. 70; and in Request for Advisory Opinion submitted by the Sub-Regional Fisheries Commission, Advisory Opinion, 2 April 2015, ITLOS Reports 2015, p. 4, para. 120 and 216.

22 Paragraph 3.4 of the IPOA-IUU refers precisely to this latter consideration by providing, as a safeguard clause, that "Notwithstanding paragraph 3.3, certain unregulated fishing may take place in a manner which is not in violation of applicable international law, and may not require the application of measures envisaged under the International Plan of Action (IPOA)".

23 Casado, 2012, p. 4.

24 Treves, 2009, p. 135.

25 COM(2002)180 final, 28.5.2002.

26 Sobrino, 2008.

27 Regulation (EU) No 1026/2012, of the European Parliament and of the Council, of 25 October 2012, OJ No. L 316, 14.11.2012.

28 The latest development has been Regulation (EU) No 1380/2013, of the European Parliament and of the Council, of 11 December 2013, on the Common Fisheries Policy, OJ No. L 354, 28.12.2013.

29 Created by Council Regulation (CE) No 768/2005, of 26 April 2015, establishing a Community Fisheries Control Agency, OJ No. L 128, 21.5.2005.

30 See, for example, the study presented to the European Parliament entitled Illegal, unreported and unregulated fishing: sanctions in the EU, Document IP/B/PECH/IC/2013–184, July 2014, available from: www.europarl.europa.eu/RegData/etudes/

STUD/2014/529069/IPOL_STU%282014%29529069_EN.pdf, p. 26ff [accessed 15 March 2020].

31 Council Regulation (EC) No 1005/2008 of 29 September 2008, OJ No. L 286, 29.10.2008.

32 Commission Regulation (EC) No 1010/2009 of 22 October 2009, OJ No. L 280, 27.10.2009.

33 Pons, 2010.

34 Article 3 of Regulation 1005/2008.

35 Article 2.5 of Regulation 1005/2008.

36 Article 1.3 of Regulation 1005/2008.

37 Pons, 2016; Honniball, 2020.

38 In May 2019, the European Commission launched CATCH, an IT System Search that aims to digitalize the currently paper-based EU catch certification scheme as laid down in Regulation (EC) No 1005/2008.

39 Reformed by the Council Regulation (EC) No 1224/2009 of 20 November 2009 establishing a Community control system for ensuring compliance with the rules of the common fisheries policy, OJ No. L 343, 22.12.2009, and by the recent Regulation (EU) No 1380/2013 on CFP.

40 Article 4 in relation to port inspections and article 31 in relation to verifications of catch certificates, both in Regulation 1010/2009.

41 Article 9.2 of Regulation 1005/2008.

42 See Chapter 8.

43 Article 27 of Regulation 1005/2008.

44 Every year, the EU incorporates the various lists of RFMOs into a single list of IUU vessels. See the latest list in Commission Implementing Regulation (EU) No 2018/1883 of 3 December 2018, OJ No. L 308, 4.12.2018.

45 Measures provided for in article 37 of Regulation 1005/2008.

46 Article 32 of Regulation 1005/2008. See in this respect COM(2015)480 final, in particular pp. 5–9.

47 The decision shall be taken by the Council, acting by qualified majority on a proposal from the Commission [article 33(1) of Regulation 1005/2008].

48 Honniball, 2020, p. 7ff.

49 Article 31.6 (a) and (b) of Regulation 1005/2008.

50 To date, Belize, Cambodia, Comoros, Guinea, Sri Lanka, St. Vincent, and the Grenadines have been included on the list of non-cooperating third states, although Belize, Guinea, and Sri Lanka have since been removed from the list after successfully reforming their fisheries governance systems.

51 Van der Marel, 2019, p. 314.

52 Tsameny et al., 2010.

53 Auld, 2018; Churchill, 2019. Although these are unilateral trade measures, article XX of GATT allows for the establishment of general exceptions that are necessary to protect human, animal or plant life or health [article XX (b)] or relating to the conservation of exhaustible natural resources [article XX (g)], provided that such measures are not applied in a manner which would constitute a means of arbitrary or unjustifiable discrimination between countries where the same conditions prevail, or a disguised restriction on international trade (the *chapeau* provision of article XX).

54 Pons, 2010, 2016.

55 Adopted by the Committee on Fisheries at its 32nd session (C 2017/23) and approved by the FAO Conference at its 40th session (C 2017/REP).

56 Adopted by the Committee on Fisheries at its 31st Session (C 2015/23) and endorsed by FAO Council at its 150th Session (CL 150/REP).

57 FAO, 2018, p. 112.

58 However, the EU is a full member only of the FAO, not of the IMO or the ILO.

59 Swan, 2006; Molenaar, 2007.

60 By Resolution 12/2009, adopted by the FAO Conference at its 36th Session, C 2009/REP, Appendix E.

61 On 20 June 2011, the Council adopted Decision 2011/443/EU approving, on behalf of the EU, the PSMA and the related Declaration on Competition, OJ No. L 191, 22.7.2011.

62 Arenas, 2009; Casado, 2012; Pons, 2012.

63 As flag member states, port member states, coastal member states, trading or importing member states and member states that must ensure that their nationals do not support IUU fishing.

64 Report of the Technical Consultation to Draft a Legally-Binding Instrument on Port State Measures to Prevent, Deter and Eliminate Illegal, Unreported and Unregulated Fishing, Rome, 23–27 June 2008, 26–30 January 2009, 4–8 May 2009 and 24–28 August 2009, available from: www.fao.org/3/a-i1122e.pdf [accessed 15 March 2020]; as well as the Negotiating Guidelines recommended by the Commission [COM(2008)333 final] and approved by the Agriculture and Fisheries Council of 23–24 June 2008 (10590/08).

65 Pons, 2012, p. 33ff.

66 Reducing to 22 the minimum number of states parties with an aggregate number of vessels greater than 24 meters in length of not less than 3,600 vessels and fine-tuning the provisions on exemptions and certifications and surveys.

67 Council Directive 97/70/EC, of 11 December 1997, setting up a harmonized safety regime for fishing vessels of 24 meters in length and over, OJ No. L 34, 9.2.1998, the provisions of which apply both to fishing vessels flying the flag of member states and to fishing vessels flying the flag of third states operating in the internal waters or territorial seas of a member state or landing catches in a port of a member state.

68 Council Decision 2014/195/EU, of 17 February 2014, OJ No. L 106, 9.4.2014.

69 Currently, only 12 countries with an aggregate fleet of 2,400 vessels are parties to the Agreement, and these include only six EU countries (Belgium, Denmark, France, Germany, the Netherlands and Spain).

70 Held in Torremolinos, from 21 to 23 October 2019, and Coinciding with the Fourth Joint IMO/FAO Working Group on IUU Fishing and Related Matters.

71 See the Declaration available from: www.imo.org/en/About/Events/Documents/Torremolinos%20fishing%20conf/TORREMOLINOS%20DECLARATION.pdf [accessed 15 March 2020].

72 Adopted by the General Conference of the ILO on 14 June 2007.

73 It is in force for Estonia, France and Lithuania and between the end of 2020 and the beginning of 2021 will enter into force for Denmark, the Netherlands, Poland and Portugal.

74 Council Directive (EU) 2017/159 of 19 December 2016 implementing the Agreement concerning the implementation of the Work in Fishing Convention, 2007 of the ILO, concluded on 21 May 2012 between the General Confederation of Agricultural Cooperatives in the European Union (Cogeca), the European Transport Workers' Federation (ETF) and the Association of National Organisations of Fishing Enterprises in the European Union (Europêche), OJ No. L 25, 31.1.2017.

References

Arenas Meza, M., 2010. Nuevos avances en la lucha contra la pesca INDNR: el Acuerdo de la FAO sobre medidas del Estado rector del puerto destinadas a prevenir, desalentar y eliminar la pesca ilegal, no declarada y no reglamentada (2009). In Sobrino Heredia, J.M. (Dir.), *Innovación y conocimiento. IV Jornadas Iberoamericanas de Estudios Internacionales*. Madrid: Marcial Pons, 71–79.

Auld, K., 2018. Trade Measures to Prevent Illegal Fishing and the WTO: An Analysis of the Settled Faroe Islands Dispute. *World Trade Review*, 17(4), 665–692.

Casado Raigón, R., 2012. El Acuerdo de la FAO de 2009 sobre medidas del Estado rector del puerto destinadas a prevenir, desalentar y eliminar la pesca ilegal, no declarada y no reglamentada. *Noticias de la Unión Europea*, 326, 3–14.

Churchill, R., 2019. International Trade Law Aspects of Measures to Combat IUU and Unsustainable Fishing. In Caddell, R. and Molenaar, E.J. (Editors), *Strengthening International Fisheries Law in an Era of Changing Oceans*. Oxford: Hart Publishing, 319–349.

Edeson, W., 2001. The International Plan of Action on Illegal, Unreported and Unregulated Fishing: The Legal Context of a Non-Legally Binding Instrument. *The International Journal of Marine and Coastal Law*, 16, 603–623.

FAO, 2018. *The State of World Fisheries and Aquaculture 2018 – Meeting the Sustainable Development Goals*. Rome: FAO.

Honniball, A.N., 2020. What's in a Duty? EU Identification of Non-cooperating Port States and their Prescriptive Responses. *The International Journal of Marine and Coastal Law*, 35, 1–36.

Molenaar, E.J., 2007. Port State Jurisdiction: Towards Comprehensive Mandatory and Global Coverage. *Ocean Development and International Law*, 38, 225–257.

Pons Rafols, X., 2010. *Implicaciones jurídicas y prácticas de la nueva normativa europea para prevenir, desalentar y eliminar la pesca ilegal, no declarada y no reglamentada*, FAO Legal Papers on Line, No. 85. Available from: www.fao.org/legal/prs-ol/lpo85-es.pdf [Accessed 15 March 2020].

Pons Rafols, X., 2012. La Unión Europea y el Acuerdo de la FAO sobre las medidas del Estado rector del puerto destinada a prevenir, desalentar y eliminar la pesca ilegal, no declarada y no reglamentada. *Revista General de Derecho Europeo*, 27, 1–40.

Pons Rafols, X., 2016. El protagonismo de la Unión Europea en la lucha contra la pesca ilegal, no declarada y no reglamentada. In Pueyo Losa, J. and Jorge Urbina, J. (Editors), *La gobernanza marítima europea. Retos planteados por la reforma de la política pesquera común*. Cizur Menor: Thomson Reuters Aranzadi, 177–201.

Riddle, K.W., 2006. Illegal, Unreported and Unregulated Fishing: Is International Cooperation Contagious? *Ocean Development & International Law*, 37(3–4), 265–297.

Rodrigo, A.J., 2019. The Disruptive Effects of Failed States Over the Legal Order of Seas and Oceans. In Oanta, G.A. (Editor), *Law of the Sea and Vulnerable Persons and Groups*. Napoli: Scientifica, Napoli, 151–169.

Serdy, A., 2017. *Pacta Tertiis* and Regional Fisheries Management Mechanisms: The IUU Fishing Concept as an Illegitimate Short-Cut to a Legitimate Goal. *Ocean Development & International Law*, 48(3–4), 345–364.

Sobrino Heredia, J.M., 2008. La reforma de la Política Pesquera Común y la pesca ilegal, no declarada y no reglamentada. *Noticias de la Unión Europea*, 277, 79–92.

Stolsvik, G., 2019. The Development of the Fisheries Crime Concept and Processes to Address it in the International Arena. *Marine Policy*, 105, 123–128.

Swan, J., 2006. Port State Measure to Combat IU Fishing: International and Regional Developments. *Sustainable Development Law and Policy*, 7(1), 38–42.

Theilen, J.T., 2013. What's in a Name? The Illegality of Illegal, Unreported and Unregulated Fishing. *The International Journal of Marine and Coastal Law*, 28, 533–550.

Treves, T., 2009. La pesca ilegal, no declarada y no reglamentada: Estado del pabellón, Estado costero y Estado del Puerto. In Pueyo Losa, J. and Jorge Urbina, J. (coord), *La cooperación internacional en la ordenación de los mares y océanos*. Madrid: Iustel, 135–158.

Tsameny, M., Palma, M.A., Milligan, B. and Mfodwo, K., 2010. The European Council Regulation on Illegal, Unreported and Unregulated Fishing: An International Fisheries Law Perspective. *The International Journal of Marine and Coastal Law*, 25(1), 5–31.

Van der Marel, E.R., 2019. Problems and Progress in Combating IUU Fishing. In Caddell, R. and Molenaar, E.J. (Editors), *Strengthening International Fisheries Law in an Era of Changing Oceans*. Oxford: Hart Publishing, 291–318.

6 The international dimension of the EU on access to justice in environmental matters

Alexandre Peñalver-Cabré

1. Introductory remarks

This chapter examines the role of the European Union (EU) at the international level in strengthening access to justice in environmental matters. It then looks at the limitations and gaps in the EU regulatory framework in this area at the EU and member state levels. It also shows how these circumstances weaken the EU's role in the international arena, in particular, before the European Union Court of Justice (EUCJ) in the wake of the case ACCC/C/2008/32 of Aarhus Convention Compliance Committee v EU, and it concludes with some final remarks.

2. The role of the EU in strengthening access to justice in environmental matters under international law

Access to justice is an indispensable mechanism for protecting the human right to the environment and for ensuring environmental enforcement. It also strengthens environmental democracy and makes sustainable development and the protection of other environment-related human rights more viable.[1]

Like many other international standards, access to justice in environmental matters has initially taken the form of soft law standards and subsequently matured into hard law standards.[2] The human right to the environment was first recognized in principle one of the 1972 Stockholm Declaration on the Human Environment[3] and subsequently in principle ten of the 1992 Rio de Janeiro Declaration on Environment and Development,[4] which specified that effective access to justice should be provided to protect the environment. However, access to justice in environmental matters has also taken shape in a number of regional international treaties protecting the human right to the environment.[5] For instance, we can mention the African Charter on Human and Peoples' Rights of 1981, which went into force on 1986, specifically articles 24 and 26; the United Nations Economic Commission for Europe (UNECE) Convention on Access to Information, Public Participation in Decision-making and Access to Justice in Environmental Matters[6] (Aarhus Convention) of 1998, which went into force in 2001, specifically articles 1 and 9; the Arab Charter on Human Rights of 2004, which went into force in 2008, specifically articles 1, 12, and 38; and the Regional Agreement of the Economic

Commission for Latin America and the Caribbean on Access to Information, Public Participation and Access to Justice in Environmental Matters of 2018, which is yet to come into force, specifically articles 4 and 8. This regional approach has been encouraged by Resolution 66/288 adopted by the United Nations (UN) General Assembly on 27 July 2012, which refers to "[t]he future we want" by "encouraging action at the regional, national, subnational and local levels to promote access to information, public participation and access to justice in environmental matters, as appropriate" (paragraph 88).[7]

In addition to these efforts, however, the adoption of a global approach to access to justice in environmental matters is becoming more and more pronounced.[8] The recognition of access to justice in environmental matters in a global international environmental treaty is now even being considered pursuant to Resolution 72/277 adopted by the UN General Assembly on 10 May 2018 entitled "Towards a Global Compact for the Environment".[9] This resolution has given a major impetus to the draft Global Pact for the Environment of 2017, which was prepared by an international group of experts and includes, precisely, the human right to the environment (article 1) and access to justice (article 11). The adoption of the Global Pact in 2022 has even been given consideration within the framework of the Earth Summit, but its success remains uncertain in light of the antagonistic positions that currently exist.

This global vision of access to justice in environmental matters has been reinforced by Resolution 70/1 adopted by the UN General Assembly on 25 September 2015, entitled "Transforming Our World: the 2030 Agenda for Sustainable Development". While access to justice in environmental matters can be inferred from various Sustainable Development Goals (SDGs), we highlight the two that are most directly related: "Promote the rule of law at the national and international levels and ensure equal access to justice for all" (16.3) and "Promote and enforce non-discriminatory laws and policies for sustainable development" (16.b).[10]

Without doubt, however, the international treaty that occupies a central place, owing to its content and its development, is the Aarhus Convention.[11] As Sommermann has pointed out, many countries regard access to justice as "the greatest transformative potential" of the Aarhus Convention.[12] The explanatory memorandum makes clear the dual purpose of access to justice: "[E]ffective judicial mechanisms should be accessible to the public, including organizations, so that its legitimate interests are protected and the law is enforced". Specifically, article 1 mentions access to justice as one of the three mechanisms to protect the human right to the environment, while article 9 contains a regulation of this mechanism, whose main features will be summarized in the following.[13]

Firstly, the three modalities of access to justice are established: a) to access environmental information by requiring that prior recourse to the judicial process be expeditious and free of charge or inexpensive (article 9.1); b) to challenge the legality (both substantive and procedural) of any administrative decision, act, or omission within the scope of article 6 (authorizations of activities mentioned in Annex I) and, if provided for by domestic law, other provisions of the Convention (article 9.2); and c) to enforce environmental legislation against actions or omissions by individuals or public authorities that commit any breach of environmental legislation not foreseen by the other two review procedures (article 9.3).[14] In these three modalities, the following appeals are provided for: a pre-court appeal and a judicial remedy (article 9.1); a judicial remedy and, optionally, a pre-court appeal (article 9.2); and a pre-court appeal or judicial remedy (article. 9.3).

Secondly, the capacity of natural and legal persons and, in cases provided for by domestic law, of groups without legal personality is recognized (article 2.4). Also, wide standing is established according to the type of access to justice: a) the public (article 9.1); b) the public concerned (article 9.2) where a sufficient interest or the violation of a right is claimed, referring to domestic law for its definition, but also prescribing a wide interpretation of access to justice and the recognition of environmental NGOs as holders of a sufficient interest in the environment (articles 2.5 and 9.2); and c) the public (article. 9.3), although the reference to domestic law allows any modality of objective standing (popular action and legal authorization) or subjective standing (sufficient interest and individual right or interest).[15]

Thirdly, important general requirements are established for all three procedures mentioned here as follows: remedies must be adequate and effective (including injunctive relief as appropriate); procedures must be fair, equitable, timely, and not prohibitively expensive (article 9.4); and there must be appropriate assistance mechanisms to remove or reduce financial and other barriers to access to justice (article 9.5).

It is true that the regulation of access to justice in the Aarhus Convention is the thinnest and most general of the three pillars. It also contains general obligations and references to national law that give flexibility to the parties for their implementation. However, in no case can they be ignored.[16] Article 27 of the Vienna Convention on the Law of Treaties of 23 May 1969 provides that "a party might not invoke the provisions of its internal law as justification for its failure to perform a treaty". Likewise, according to article 31.1 of the Vienna Convention, a treaty must be interpreted in good faith in accordance with the ordinary meaning to be given to the terms of the treaty in their context and taking into account its object and purpose, which in this case can be summed up in one main purpose, namely effective access to justice. The Aarhus Convention marks out a clear horizon in access to justice that requires a transformation of the law of the parties to articulate effective mechanisms of environmental protection. As we shall see, the EU can certainly be said to have played a significant role in promoting the inclusion of access to justice in international law at both international and EU levels.

At the international level, the EU participated in the Rio Conference, which resulted in the aforementioned Principle 10 of the Rio Declaration. More importantly, however, the EU played a decisive role in the birth of the Aarhus Convention and it is the only regional international economic integration organization that has been a party to the Convention since its ratification by Council Decision 2005/370/EC of 17 February 2005 on the conclusion, on behalf of the European Community, of the Aarhus Convention.[17] Furthermore, the EU has made clear its active support for the Convention and its implementation at the regional level of the UNECE, as witnessed by the EU Declaration upon signature of the Aarhus Convention:

> The European Community wishes to express its great satisfaction with the present Convention as an essential step forward in further encouraging and supporting public awareness in the field of environment and better implementation of environmental legislation in the UNECE region, in accordance with the principle of sustainable development. Fully supporting the objectives pursued by the Convention and considering that the European Community itself is being actively involved in the protection of the environment.

At the EU level, the EU has also adopted legislation and case law pursuant to the Aarhus Convention, as we shall see in the next section, without prejudice to the fact that it has also highlighted the following existing bases of EU primary law for their adoption: a) sustainable development as one of the main principles of the EU (article 3.3 TEU), b) the fundamental rights to effective judicial protection and to free legal aid for those who lack sufficient resources (articles 19.1 TEU and 47 CFREU), and c) environmental protection in the EU Charter of Fundamental Rights (article 37).[18] However, it should be noted that the main purpose of access to environmental justice in the EU has not been so much concerned with the protection of the human right to the environment as to ensure the enforcement of environmental law.[19]

In addition, the EU has taken steps to adopt non-legislative measures to ensure awareness and enforcement of access to justice in environmental matters by legal operators and the public, in accordance with articles 3.2 and 9.5 of the Aarhus Convention. A good example is the European General Framework for Judicial Training to create EU law training material for member states' judges, prosecutors, and judicial staff on the application of EU legislation, including access to justice in environmental matters.[20] However, it is also worth mentioning the "Citizen's Guide to Access to Justice in Environmental Matters" produced by the European Commission Directorate-General for Environment in 2018, which is aimed at the general public.

Finally, the EU has shown interest in strengthening access to justice in environmental matters through the development of the SDGs. In this sense, the EU Commission Communication on Next steps for a sustainable European future,[21] emphasizes that "the Commission will strengthen the tracking of progress of environmental objectives through the Environmental Implementation Review and will launch initiatives to . . . facilitate access to justice, and support environmental compliance in Member States (SDG 17)".

3. The EU's inadequate implementation of the Aarhus Convention on access to justice in environmental matters

Although, as we have seen, the EU played an important role in the emergence of the Aarhus Convention, we must nonetheless examine whether the EU's subsequent implementation of the Convention has been adequate. This is very important because 60% of the Aarhus state parties are EU member states, and because the EU is a multilevel system (EU and member states).[22] The Aarhus Convention calls upon the EU to adopt its law on access to justice in environmental matters, which has taken place in varying ways depending on whether addressed to EU member states or to EU institutions and bodies.[23] In the analysis of this dual implementation by addressee, we can highlight a differential treatment that is temporal in nature (the timing of the approval of legislation) and material in nature (content of legislation).

If we look first at the timing of the adoption of rules, the EU regulations for member states were mainly adopted before the ratification of the Convention, without prejudice to any subsequent updates that might be necessary. This is acknowledged by the EU Declaration upon approval:

> [T]he European Community declares that it has already adopted several legal instruments, binding on its Member States, implementing provisions of this Convention and will submit and update as appropriate a list of those legal instruments to the Depositary in accordance with Article 10 (2) and Article 19 (5) of the Convention.

The regulations passed before ratification of the Aarhus Convention were limited to environmental information, integrated pollution control and environmental impact assessments.[24] After ratification, the Convention was reproduced in new legislation on integrated pollution control, environmental impact assessments, and environmental liability.[25]

However, upon approval, the aforementioned EU Declaration noted that these regulations did not fully comply with the obligations of article 9.3 of the Aarhus Convention because it considered that their approval was a matter for member states, since they involved appeals against private persons and public authorities other than EU institutions and bodies. It also warned, however, that this would not prevent the EU from approving regulations aimed at member states to comply with the obligations of the article. We see, therefore, that the EU recognized considerable deference towards member states in the implementation of article 9.3, which corresponds, to a large extent, to the role of the state courts as ordinary courts in the application of EU environmental law, without prejudice to the fact that they can pose preliminary rulings to the European Union Court of Justice (EUCJ).

This stands in contrast to the European Commission's position, which sought the adoption of general rules on access to justice in environmental matters. The European Commission warned of its intention in the Statement annexed to Directive 2003/35/EC[26] and subsequently submitted the Proposal for a Directive of the European Parliament and of the Council on access to justice in environmental matters.[27] However, further progress on the Directive was blocked by the opposition of several states on the Council. After a decade, the European Commission redoubled its efforts to break the deadlock by commissioning various studies[28] and obtaining the support of the European Parliament and the Committee of the

Regions.[29] In the end, however, the persistent opposition of some states led the European Commission to withdraw its proposal for a directive in 2014 and it instead adopted the EU Commission Notice: Access to justice in environmental matters.[30] The main purpose of the Notice is to collect and systematize the case law of the EUCJ, which is normally based on preliminary rulings from state courts, but it also makes novel contributions to the interpretation of the Aarhus Convention and EU law on access to justice in environmental matters. Thus, there has been a recognition of the EUCJ's central role in the implementation of article 9.3 Aarhus Convention by member states.[31]

It should be noted, however, that the regulations concerning EU institutions and bodies were adopted after the ratification of the Convention because it was felt that they should be applied in accordance with existing EU law, without prejudice to possible future minor amendments. In this respect, the EU Declaration upon approval states that

> the Community reiterates its declaration made upon signing the Convention that the Community institutions will apply the Convention within the framework of their existing and future rules on access to documents and other relevant rules of Community law in the field covered by the Convention.

However, having found that EU law was insufficient,[32] the EU passed Regulation 1367/2006 (the so-called "Aarhus Regulation") on 6 September 2006, regarding the application of the provisions of the Aarhus Convention on Access to Information, Public Participation in Decision-making and Access to Justice in Environmental Matters to Community institutions and bodies.

Beyond the differential treatment shown in the timing of the approval of rules, it is perhaps even more important to examine the content of the EU regulations implementing the Aarhus Convention. In general, there has been a very weak development of the Aarhus Convention, which is often limited to reproducing what the Convention already provides. At times, the reproduction is even incomplete and important obligations are forgotten.[33] This is a further example of the governance challenges in international environmental law, which Sands has characterized as "the manifestly inadequate implementation of many obligations at the domestic level in many states".[34] In spite of this limited content, however, it is also important to highlight the contrast between the greater demands put on EU member states and the lower demands put on EU institutions and bodies, as we will analyse in the following.

With regard to the regulations on access to justice in environmental matters addressed to member states, the EU has limited itself to approving some sectoral environmental directives that contain minimum regulations on access to justice, basically reproducing the obligations of the Convention.[35] The EU could have gone further to develop the Aarhus Convention, given its consideration as a mixed international treaty, in the context of a shared EU competence such as the environment (article 4.2.e TFEU). The main limitations of the regulations that have been adopted are shown in the following.

The appeal of the public to obtain environmental information (article 9.1 Aarhus Convention) is provided for in article 6 of Directive 2003/4/EC. However, we note three main shortcomings: a) it does not establish the binding nature of the resolutions of pre-court appeals; b) it does not include the important requirements in articles 9.4 and 9.5 (adequate and effective remedies; procedures that are fair, equitable, timely, and not prohibitively expensive; and appropriate assistance mechanisms to remove or reduce financial and other barriers to access to justice), only the requirements in article 9.1 for preliminary non-judicial proceedings to be expeditious and free of charge or inexpensive; and c) the non-judicial proceedings should have been reinforced much more by means of independent bodies to achieve an effective protection of access to environmental information.[36]

Appeals by the public to challenge the legality of any administrative decision, act, or omission (article 9.2) are provided for in the aforementioned environmental Directives on integrated pollution control and environmental impact assessments. These Directives reproduce the standing of the public concerned.[37] They also reiterate many of the requirements in article 9.4 of the Aarhus Convention stipulating that remedies must be fair (rather than objective), equitable, expeditious, and not unduly burdensome, but they omit adequate and effective remedies (including injunctive relief as appropriate). The requirements in article 9.5 of the Aarhus Convention are also missing.[38]

The principal shortcoming, however, relates to the public appeal procedure laid out in article 9.3 of the Aarhus Convention, because public appeals are only provided for in the Directive on environmental liability, which follows the reviewing and subjective model for appeals in article 9.2 of the Aarhus Convention. Furthermore, there is no reference to the general requirements laid out in articles 9.4 and 9.5. As we have pointed out, this limited scope is due to the impossibility of adopting a Directive on access to justice in environmental matters, despite the Commission's attempts.

The lack of regulatory development of the appeals procedure in article 9.3 and the emergence of unforeseen aspects of the appeals procedures in articles 9.1 and 9.2 could be understood as having been remedied by the case law of the EUCJ, largely contained in EU Commission Notice 2017/C 275/01, on practically all aspects of access to justice (standing, scope of judicial review, effective remedies, cost, time limits, timelines, and the efficiency of procedures). The EUCJ has been requiring state courts to interpret the rules on access to justice in environmental matters in accordance with the Aarhus Convention, as stated in the judgment of the EUCJ of 8.3.2011 in the *Slovak Brown Bear* case where, despite not recognizing the direct effect of article 9.3 of the Aarhus Convention, an obligation was established for state courts to interpret the rules in accordance with the Convention.[39]

On the other hand, the EU has put lower demands on EU institutions and bodies, both at the normative and jurisprudential levels, in their compliance with the obligations of the Aarhus Convention on Access to Justice in Environmental Matters. As IMPEL has pointed out: "There is a certain irony in the difference between the standards which the Community obliges the Member States to apply . . . and those to which the Community is itself committed".[40]

Certainly, the Aarhus Regulation has sought to overcome the clear individualistic limitations of appeals for annulment and omission to the EUCJ (articles 263–266 TFEU) insofar as they restrict the standing of natural or legal persons to some of these three cases: a) act addressed to that person; b) act which is of direct and individual concern to the person; and c) a regulatory act which is of direct concern to the person and does not entail implementing measures.[41] However, the Aarhus Regulation has proved unable to overcome these limitations, as we shall see in the following.[42]

In relation to the public's right of access to environmental information under article 9.1 of the Aarhus Convention, article 3 of the Aarhus Regulation refers to the remedies provided for in articles 7 and 8 of Regulation (EC) No 1049/2001 of the European Parliament and of the Council of 30 May 2001 regarding public access to European Parliament, Council, and Commission documents. These provisions provide for an administrative appeal that may be lodged with the same institution or body and, in the event of refusal, an action for annulment with the EUCJ (article 263 TFEU) and/or a complaint with the European Ombudsman (article 228 TFEU). We insist that the mere provision of a decision by the same institution, without any guarantee of impartiality and independence, does not meet the requirements of effectiveness, objectivity, and fairness.

Furthermore, the only appeal is provided for in articles 10–12 of the Aarhus Regulation, based on article 9.3 of the Aarhus Convention. Articles 10–12 provide for the objective standing of NGOs that meet certain requirements (non-profit-making legal persons, primary objective of environmental protection, two years' seniority, and actively working to achieve their aims) to "make a request for internal review to the Community institution or body that has adopted an administrative act under environmental law or, in case of an alleged administrative omission, should have adopted such an act". Subsequently, they may institute proceedings before the EUCJ in accordance with the relevant provisions of the Treaty. Let us look at the major limitations of this appeals procedure in relation to the provisions of the Aarhus Convention.

Firstly, the appeals procedure does not recognize the capacity of groups without legal personality that is allowed by the Aarhus Convention and that some EU and EUCJ rules sometimes admit in other non-environmental areas.[43] It neither provides for the standing of natural persons that appears in the Aarhus Convention (articles 9.2 and 9.3), nor does it provide for the standing of environmental NGOs that is characteristic of the appeal of the public concerned (article 9.2 Aarhus Convention).

Secondly, the appeals procedure is restricted to certain administrative acts and omissions that meet the following requirements: a) they are of individual scope, which is much more restrictive than the actions or omissions of general or collective scope referred to in the application of environmental legislation; and b) they must have a legally binding and external effect. Undoubtedly, the individualist legitimation of the appeal for annulment under article 263 of the TFEU[44] and its restrictive interpretation by the EUCJ have been criticized by a wide range of legal scholars alleging that a more favourable interpretation of environmental interests is possible.[45] In fact, the General Court has tried to eliminate such an individualistic conception based on the interpretative effectiveness of article 9.3 of the Aarhus Convention on the TFEU and the Aarhus Regulation. However, this has been wholly rejected by the EUCJ (Grand Chamber), which has denied any interpretative effectiveness of article 9.3 of the Aarhus Convention that would overcome the individualistic configuration of the appeal for annulment in the TFEU and the Aarhus Regulation.[46] We see here how the EUCJ shows the absolute precedence of the TFEU over article 9.3 of the Aarhus Convention, in contradiction of its own more incisive jurisprudence towards EU member states which requires an interpretation of the rules on access to justice that complies with the Aarhus Convention.

The appeals procedure also contains other limitations on administrative acts and omissions contrary to article 9.3 of the Aarhus Convention: a) they must be "under environmental law" instead of "relating to the environment"; b) they can involve violations only by public authorities, not by private individuals, a limitation which is justified in par. 18 by requirements of compatibility with the Treaty without specifying what these requirements are; and c) the procedure excludes cases in which the EU body acts as a review body, citing, among other things, the procedure for non-compliance with European law by states.

Thirdly, the important general requirements set out in articles 9.4 and 9.5 of the Aarhus Convention, which we have highlighted on several occasions throughout this chapter, are neither cited nor developed.

Finally, the divergence in the greater demand put on EU member states than on EU institutions and bodies is also evident on a practical level, specifically in relation to the information provided to citizens on their access to justice on the environment. For instance, the European Commission published the "Citizen's Guide to Access to Justice in Environmental Matters" (2018), which is limited to access to justice from member states. However, there is no equivalent guide on access to the EUCJ in environmental matters.[47]

4. The Aarhus Convention Compliance Committee case ACCC/C/2008/32 and the shortcomings in access to justice in environmental matters at the EU level

The inadequacies of the system of access to justice before the EUCJ in environmental matters arising from the limitations of EU law and the restrictive interpretations of the EUCJ were the subject of the communication presented by the NGO Client Earth on 1 December 2008 before the Aarhus Convention Compliance Committee (ACCC/C/2008/32). In particular, the communication focused on the lack of standing of natural persons, the lack of standing in the interest of environmental NGOs, the individual scope of the acts challenged, the fact that the EUCJ's interpretation was much more favourable to economic interests than to environmental public interests, and the prohibitive costs.

The Compliance Committee issued a first finding on 14 April 2011 (EU ACCC/C/2008/32, Part I) in which, awaiting certain judgments still pending from the EUCJ (Grand Chamber), it advances that the EU violates articles 9.3 and 9.4 of the Aarhus Convention if the case law of the EUCJ is upheld or no new regulations are adopted to take into account the obligations of the Aarhus Convention on the standing of natural persons and the standing in the interest of environmental protection NGOs. On the other hand, it rejects the violation of articles 9.4 and 9.5 in relation to prohibitive costs, finding that the allegations regarding the lack of necessary resources were not sufficiently substantiated.

Subsequent to Judgments of the Court (Grand Chamber) which rejected any interpretative effectiveness of article 9.3 of the Aarhus Convention to overcome the individualistic restrictive interpretation of access to justice in environmental matters, the Compliance Committee issued a second finding on 17 March 2017 (EU ACCC/C/2008/32, Part II) stating, now with full force and effect, that the EU has failed to implement the Aarhus Convention. This finding states that the EU has violated articles 9.3 and 9.4 of the Aarhus Convention because neither the jurisprudence of the EUCJ (there has been no new direction in the jurisprudence) nor the Aarhus Regulation (the Aarhus Regulation does not correct or compensate for the failings in the jurisprudence) implements or complies with the obligations arising under those paragraphs of article 9.

The Compliance Committee criticizes the fact that the EUCJ requires an interpretation from state courts in accordance with article 9.3 of the Aarhus Convention, but does not apply it to itself. It recommends that the Aarhus Regulation be amended or new rules adopted to comply with articles 9.3 and 9.4, and that the EUCJ interpret EU law to the fullest extent possible in accordance with the objectives of these articles. In particular, it proposes the following changes: a) to include members of the public other than NGOs, b) to extend to general acts and not only acts of individual scope, c) to extend the acts subject to appeal to all those relating to the environment and not only under environmental law, and d) to permit appeals against acts that do not have legally binding and external effects.[48]

5. EU response to the meeting of the parties of Aarhus Convention in case ACCC/C/2008/32

In accordance with Decision I/7 MOP review of compliance ECE/MP.PP/2/Add.8, 2 April 2004, the findings of the Compliance Committee were included in draft Decision VI/8f, regarding the compliance of the EU with its obligations under the Aarhus Convention, which was submitted to the 6th Meeting of the Parties in September 2017.

The European Commission made every effort to get the EU to vote against the draft Decision because it considered that the draft Decision called into question essential constitutional principles of EU law and that any shortcomings in public access to the EUCJ could be compensated for by the preliminary ruling (article 267 TFEU).[49] It should be also noted that the Compliance Committee (ACCC/C/2008/32, Part I, par. 90) had already stated that the preliminary ruling alone did not comply with Aarhus because it is not a direct protective mechanism before the EUCJ and its approach is conditional on the decision of the state court.[50]

By contrast, the Council chose to reaffirm its full support for the important objectives of the Aarhus Convention and to accept draft Decision VI/8f, but to propose some modifications that seriously brought into question the effectiveness of the Aarhus Convention's compliance mechanisms.[51] Firstly, the Council sought to mitigate the effects of the resolution by eliminating any legally binding effects through the replacement of "endorses" with "takes note" and the insertion of the words "to consider" after the words "Recommends to the Party concerned". Secondly, the Council sought to remove references to both the failure of EUCJ case law to implement or comply with Aarhus Convention obligations and to any reliance on possible changes in future case law. The Council justified these changes on the basis of the reasons given by the European Commission already explained. Furthermore, the Council made it clear that the separation of powers within the Union prevented the Council from giving instructions or making recommendations to the EUCJ to change its case law. It therefore suggested that the solution would be a regulatory change that should be undertaken in accordance with the fundamental principles of EU law.

This EU position raised a serious challenge to the compliance mechanism of the Aarhus Convention and it was not welcomed at the Meeting of the Parties, drawing sharp criticism from some states, NGOs, and members of the Compliance Committee. It was argued that the EU, like all other parties, had to comply with the obligations of the Convention without special treatment and that the courts, as organs of states, should also comply with international treaties, as in the case of the EU in article 216.2 of the TFEU.[52] The EU's lack of respect for the Aarhus Convention is well reflected in the minutes of the Meeting of the Parties: "Faced with a situation that could seriously jeopardize the authority of the Meeting of the Parties and the integrity of the Convention's compliance mechanism" (ECE/MP.PP/2017/2). Therefore, it was agreed, in order to reach consensus and considering the exceptional circumstances, to postpone the decision-making on draft Decision VI/8f to the next ordinary session of the Meeting of the Parties to be held in 2021. However, the EU again recorded its "willingness to continue exploring ways and means to comply with the Convention in a way that was compatible with the fundamental principles of the European Union legal order and its system of judicial review".[53]

While Parliament has clearly advocated an amendment to the Aarhus Regulation, the Council has been more cautious. Issuing Council Decision (EU) 2018/881 of 18 June 2018, the Council requested the Commission to submit, by 30 September 2019, a study on the Union's options for addressing the findings, and to submit, by 30 September 2020, if appropriate in view of the outcomes of the study, a proposal for amending the Aarhus Regulation, or otherwise to inform the Council on other measures.

The Commission's first action was to carry out a public consultation between 20 December and 14 March 2019, and then to order a study, which set out various measures that can be grouped into three types: (a) maintaining the current situation, (b) adopting guidelines or interpretative communications, and (c) amending the Aarhus Regulation and/or adopting a directive on access to justice in environmental matters.[54] After completion of the study, the Commission insisted on recalling, once again, the need to abide by the fundamental principles of the EU legal order and its system of judicial review. However, it has also become more open to using its monopoly on regulatory initiative to propose amendments to the Aarhus Regulation and to put forward other regulatory initiatives to access the EUCJ. However, the European Commission also recalled that the Council has prevented the adoption of legislation proposed by the European Commission on access to justice in environmental matters and warned that it intended to take action again to improve access to state courts as well.[55] This was confirmed in the Communication from the Commission on the European Green Deal,[56] which states:

> The Commission will consider revising the Aarhus Regulation to improve access to administrative and judicial review at EU level for citizens and NGOs who have concerns about the legality of decisions with effects on the environment. The Commission will also take action to improve their access to justice before national courts in all Member States.

Surprisingly, however, these legislative reforms are not included in the key actions set out in the Annex to the Communication on the European Green Deal Roadmap.

In sum, a static interpretation of the fundamental principles of EU law (in particular, the traditional system of access to the EUCJ) and the divergences between the Council and the European Commission on how to correctly implement the Aarhus Convention on access to justice (in particular, before the EUCJ) have had two main negative consequences: an inadequate implementation of the Aarhus Convention by the EU and a weakening of the role of the EU before the Aarhus Convention enforcement bodies.

6. Final remarks

The Aarhus Convention marks out a clear horizon in access to justice that calls for a profound transformation of the law of the EU and its member states to articulate effective mechanisms of environmental protection. Although the EU played an important role in the emergence of the Aarhus Convention, it is also necessary to examine its subsequent implementation, which takes on twofold international significance given that 60% of the state parties are members of the EU and that the EU is a multilevel system (EU and member states). Certainly, the EU has adopted regulations on access to justice in environmental matters, but it has done so in different ways depending on whether the regulations are addressed to EU member states or to EU institutions and bodies.

Firstly, from the viewpoint of timing, the regulations for member states were mainly approved before the ratification of the Convention, whereas the regulations for EU institutions and bodies have been approved since ratification because it was considered that the Convention should be applied in accordance with existing EU law, without prejudice to possible future rules, such as the Aarhus Regulation.

Secondly, from the viewpoint of the content of regulations, there has been a very weak development of the Aarhus Convention by the EU, which has limited itself to reproducing, in a partial way, what was already provided by the Convention. However, we can observe a greater demand on EU member states in contrast to a lower demand on EU institutions and bodies.

At the state level, the EU has approved some sectoral environmental Directives that contain minimal and insufficient regulations on access to justice. The EU could have gone further to develop the Aarhus Convention because of the Convention's status as a mixed international treaty. However, this limited scope is due to the impossibility of approving a directive on access to justice in environmental matters, despite the attempts of the European Commission. Nevertheless, the position of the EUCJ in favour of the requirements of the Aarhus Convention with regard to member states has partly compensated for this normative shortcoming by converting jurisprudence into an important actor of implementation.

On the other hand, the EU has put a lower demand on EU institutions and bodies, both at regulatory and at jurisprudential levels, in their compliance with the obligations of the Aarhus Convention on Access to Justice in Environmental Matters. The attempts of the General Court to eliminate the individualistic conception of the appeal for annulment and omission of the TFEU and the Aarhus Regulation have been totally rejected by the EUCJ (Grand Chamber), which denies any interpretative effectiveness of article 9.3 of the Aarhus Convention that would allow this individualistic configuration to be overcome. The EUCJ (Grand Chamber) shows an absolute precedence of the TFEU over article 9.3 of the Aarhus Convention, which contradicts its own more demanding jurisprudence towards member states. The Compliance Committee Case (ACCC/C/2008/32) has therefore declared the system of access to justice before the EUCJ to be contrary to articles 9.3 and 9.4 of the Aarhus Convention, criticizing the fact that the EUCJ requires state courts to interpret these provisions, but does not apply them to itself.

Finally, the traditional system of access to the EUCJ and the divergences between the Council and the European Commission on how to correctly implement the Aarhus Convention in the field of access to justice (in particular, before the EUCJ) have had two main negative consequences: an inadequate implementation of the Aarhus Convention in the EU and a weakening of the role of the EU before the Aarhus Convention enforcement bodies.

Notes

1 These connections appear in various works by the Special Rapporteurs on Human Rights and the Environment (John Knox and, currently, David R. Boyd) that the UN Human Rights Council began to appoint from 2012. See Kravchenko and Bonine, 2008, pp. 311–366; Collins, 2015, pp. 219–244; Sands et al., 2018, pp. 217–229, 811–828.
2 Juste Ruiz and Castillo Daudí, 2014, pp. 42–46.
3 *Report of the United Nations Conference on the Human Environment*, UN Doc.A/CONF.48/14, at 2 and Corr.1 (1972).

4 *Report of the United Nations Conference on Environment and Development*, UN Doc. A/CONF.151/26 (Vol. I), 12 August 1992, Annex I.

5 Boyd, 2012; Sec and Jendrośka, 2019, pp. 533–545.

6 UNTS, vol. 2161, p. 447.

7 A/RES/66/288.

8 United Nations Environment Programme, 2010 (see Guidelines 15–26 on access to justice).

9 A/RES/72/277.

10 A/RES/70/1. See Chapter 9.

11 Wates, 2005, pp. 2–11; Beyerlin, 2015, pp. 333–339, 352; Hey, 2015, pp. 353–362; Peñalver, 2016, pp. 208–219; Sommermann, 2017, pp. 323–328, 337; Sands et al., 2018, pp. 95, 827, 931; Salazar Ortuño, 2019, pp. 28–88.

12 Sommermann, 2017, p. 326.

13 Sec (ed.), 2003, pp. 7–69.

14 This residual nature of the appeal laid out in article 9.3 has been recalled by the UNECE, 2014, pp. 197, 206 and various findings of the Compliance Committee, such as that of 28 July 2006 (Belgium ACCC/C/2005/11, para. 26).

15 Bonine, 2003, pp. 32–37; García Ureta, 2005, pp. 64–74; Jendrośka, 2005, pp. 19–20; 2006, pp. 81–82; Razquín Lizarraga and Ruíz de Apodaca, 2007, p. 364; Darpö, 2013, pp. 28–29; Jans, 2013, p. 156.

16 Decision III/3 of the Meeting of the Parties to the Aarhus Convention on promoting effective access to justice ECE/MP.PP/2008/2/Add.5.

17 OJ No. L 164M, 16.6.2006.

18 UNECE, 2014, p. 168.

19 Decision No 1386/2013/EU of the European Parliament and of the Council of 20 November 2013 on a General Union Environment Action Programme to 2020 "Living well, within the limits of our planet", OJ No. L 354, 28.12.2013. Priority objective 4 is "[t]o maximise the benefits of Union environment legislation by improving implementation" and one of its key area is: "Union citizens will have effective access to justice in environmental matters and effective legal protection, in line with the Aarhus Convention and developments brought about by the entry into force of the Lisbon Treaty and recent case law of the Court of Justice of the European Union" (point 62). Also, see EU Commission Communication: EU law: Better results through better application, (2017/C 18/02), pp. 10–13 and EU Commission Notice: Access to justice in environmental matters, 28.04.2017, (2017/C 275/01), par. 2, 12, 16, 18 and 31 to 57. See Krämer, 1997, p. 316, 2012, pp. 414, 416, 2014, pp. 247, 252; Peñalver i Cabré, 2008, pp. 352–355, 2016, pp. 220–229, 2019, pp. 126–129.

20 This programme is mentioned by the UNECE, 2014, pp. 64, 206.

21 COM(2016)739, 22.6.2016, p. 10.

22 Sommermann, 2017, pp. 328–332.

23 Hey, 2015, pp. 368–369.

24 Directive 2003/4/EC of the European Parliament and of the Council of 28.1.2003 on public access to environmental information, OJ No. L 41, 14.2.2003 (art. 6) and Directive 2003/35/EC of the European Parliament and of the Council of 26 May 2003 providing for public participation in respect of the drawing up of certain plans and programmes relating to the environment and amending with regard to public participation and access to justice Council Directives 85/337/EEC and 96/61/EC, OJ No. L 156, 25.6.2003 (art. 10bis to Directive 85/337/EEC and art. 15bis to Directive 96/61/EC).

25 Directive 2004/35/CE of the European Parliament and of the Council of 21 April 2004 on environmental liability with regard to the prevention and remedying of environmental damage, OJ No. L 143, 30.04.2004 (art. 12 and 13); Directive 2010/75/EU of the European Parliament and of the Council of 24 November 2010 on industrial emissions (integrated pollution prevention and control), OJ No. L 334, 17.12.2010 (art. 25)

and Directive 2011/92/EU of the European Parliament and of the Council of 13 December 2011 on the assessment of the effects of certain public and private projects on the environment, OJ No. L 26, 28.1.2012 (art. 11). Art. 11 of the latter Directive refers to art. 23 of Directive 2012/18/EU of the European Parliament and of the Council of 4 July 2012 on the control of major-accident hazards involving dangerous substances, amending and subsequently repealing Council Directive 96/82/EC, OJ No. L 197, 24.7.2012.

26 "Statement by the Commission: With reference to the Commission Work Programme 2003, the Commission confirms its intention to present a proposal for a directive addressing the implementation of the Aarhus Convention in respect of access to justice in environmental matters, which is envisaged for the first quarter of 2003".

27 2003/246 (COD), COM(2003)624, 24.10.2003.

28 These studies highlighted the great diversity of domestic laws and the urgency of such a Directive to establish minimum criteria for compliance with the Aarhus Convention, while respecting the principle of subsidiarity. We would like to highlight the following four: Sadeleer, Nicolas de, Roller, Gerhard and Dross, Miriam, *Access to Justice in Environmental Matters, ENV.A.3/ETU/2002/0030, Final report*, 2003; MILIEU Environmental Law and Policy, *Summary Report on the inventory of EU Member States' measures on access to justice in environmental matters*, Belgium, 2007; Darpö, Jan, *Effective Justice? (Synthesis report of the study on the Implementation of Articles 9.3 and 9.4 of the Aarhus Convention in the Member States of the European Union)*, European Commission, 2013, which is reproduced as chapter 8 of Jans, Macrory, Moreno Molina (eds.), 2013; and Faure, Michael and Philipsen, Niels (coords.), *Possible initiatives on access to justice in environmental matters and their socio-economic implications (DG ENV.A.2/ETU/2012/0009rl) Final Report,* Maastricht University (Faculty of Law, Metro), Maastricht, The Netherlands, 2013.

29 European Parliament resolution of 20 April 2012 (2011/2194(INI)) and Committee of the Regions opinion of 30 November 2012 (2013/C 17/07).

30 2017/C 275/01, 28.04.2017.

31 EU Commission Notice 2017/C 275/01 (para. 9 and 10). On the situation after the failed proposal for a directive, inter alia, Hey, 2015, pp. 369–372; Krämer, 2012, pp. 415–416, 2018, pp. 19–20; Peñalver i Cabré, 2019, pp. 124–125; Ruíz de Apodaca, 2018, pp. 6–12.

32 UNECE, 2014, pp. 76–77, 92, 194–198, 204–205.

33 Krämer, 2012, pp. 415–416, 2014, pp. 252–253; Peñalver i Cabré, 2016, pp. 229–244, 280–284; Salazar Ortuño, 2019, pp. 106–147; Martinez Jimenez, 2018.

34 Sands, 2018, p. 935.

35 On EU regulations on access to justice in environmental matters and the influence of the Aarhus Convention, among others, García Ureta, 2005, pp. 63–88; Peñalver i Cabré, 2008, pp. 352–374, 2016, pp. 207–244.

36 Nagy, 2018, p. 41; Salazar Ortuño, 2019, p. 108.

37 The public concern standing has been interpreted from a wide access to justice standard by the EUCJ. Among others, Case C-240/09, *Slovak Brown Bear* case, C-240/09, *Lesoochranárske zoskupenie VLK v Ministerstvo životného prostredia Slovenskej republiky*, 8 March 2011, ECLI:EU:C:2011:125 and case C-115/09, *Bund für Umwelt und Naturschutz Deutschland, Landesverband Nordrhein-Westfalen eV v Bezirksregierung Arnsberg*, 12 May 2011, ECLI:EU:C:2011:289. See Commission Notice 2017/C 275/01 (para. 37–40).

38 Salazar Ortuño, 2019, p. 108.

39 Case C-240/09, *Lesoochranárske zoskupenie VLK v Ministerstvo 'ivotného prostredia Slovenskej republiky*, 8 March 2011, ECLI:EU:C:2011:125.

40 IMPEL, 2000, p. 37.

41 Rehbinder and Stewart, 1985, pp. 146–149; García Ureta, 1996, pp. 113–146; Krämer, 2004, pp. 27–29; 2018, pp. 19–24; Plaza Martín, 2005, pp. 414–415, 1128–1152; Ortega Gómez and Marta, 2006, pp. 757–797.

42 Poncelet, 2012, pp. 296–309.

43 See the following definition of person in article I.4.6 *ReNEUAL* Model Rules on EU Administrative Procedure: " 'Person' means any natural or legal person. Other associations, organizations or groups may be considered as a person on the basis of sector specific EU law or the case law of the Court of Justice of the European Union" (Craig et al., 2017).

44 "Any natural or legal person may, under the conditions laid down in the first and second paragraphs, institute proceedings against an act addressed to that person or which is of direct and individual concern to them, and against a regulatory act which is of direct concern to them and does not entail implementing measures".

45 The origin of this restriction is the case 25–62, *Plaumann & Co.* v *Commission of the European Economic Community*, 15 July 1963, ECLI:EU:C:1963:17 and it is reflected in the environment (among other contexts) in case C-321/95, *Stichting Greenpeace Council (Greenpeace International) and Others* v *Commission of the European Communities*, 2 April 1998, ECLI:EU:C:1998:153. Critically, Krämer, 1997, pp. 302–317, 2012, pp. 414–416, 2014, pp. 252–253; Lenaerts and Gutiérrez-Fons, 2011, pp. 11–20, 22–24; Jans and Vedder, 2011, pp. 249–250; Jans, 2013, pp. 161–164.

46 Case C-401/12 P to C-403/12 P, *Council of the European Union and Others* v *Vereniging Milieudefensie and Stichting Stop Luchtverontreiniging Utrecht*, 13 January 2015, ECLI:EU:C:2015:4, para. 52–62. This judgment corrects the non-individualistic interpretation on case T-396/09, *Vereniging Milieudefensie and Stichting Stop Luchtverontreiniging Utrecht* v *Commission*, 13 June 2012, ECLI:EU:T:2012:301. Also, case C-404/12 P, *Council and Commission* v *Stichting Natuur en Milieu and Pesticide Action Network Europe*, set aside the case T-338/08, *Stichting Natuur en Milieu and Pesticide Action Network Europe* v *Commission*, 14 June 2012, ECLI:EU:T:2012:300. Among others, Sommermann, 2017, pp. 331–332; Martínez Jiménez, 2018, pp. 96–100; Berthier et al., 2019, pp. 58–61, 65–66.

47 According to articles 3.3 and 9.5 of the Aarhus Convention, public authorities shall ensure that information on access to justice is provided to the public. See Sec (ed.), 2003, p. 63.

48 Sands et al., 2018, pp. 149, 175–176.

49 European Commission Proposal for a Council Decision on the position to be adopted, on behalf of the European Union, at the sixth session of the Meeting of the Parties to the Aarhus Convention regarding compliance case ACCC/C/2008/32, COM(2017)366 final, 29.6.2017.

50 Krämer, 2012, pp. 415–461; Nagy, 2018, pp. 40–41; Berthier et al., 2019, pp. 67–68.

51 Council Decision (EU) 2017/1346 of 17 July 2017 on the position to be adopted, on behalf of the European Union, at the sixth session of the Meeting of the Parties to the Aarhus Convention as regards compliance case ACCC/C/2008/32.

52 Berthier et al., 2019, pp. 68–70; Salazar Ortuño, 2019, pp. 44–45, 123.

53 Report of the sixth session of the Meeting of the Parties, Budva, Montenegro, 11–13 September 2017 (ECE/MP.PP/2017/2) (point 62).

54 Milieu Consulting Sprl, *Study on EU implementation of the Aarhus Convention in the area of access to justice in environmental matters*, Final report, September 2019 07.0203/2018/786407/SER/ENV.E.4.

55 Commission Staff Document 10 October 2019 (SWD(2019) 378 final), Report on European Union implementation of the Aarhus Convention in the area of access to justice in environmental matters: "As for legislative options, the Commission has a right of initiative but only the Union legislature can adopt legislation. As noted in Section 3.4.2, the Union co-legislators have either rejected or failed to adopt several Commission proposals on environmental access to justice. In terms of future perspectives for addressing the ACCC findings, this pattern is to be noted" (p. 29).

56 COM(2019)640, 11.12.2019.

References

Berthier, A. et al., 2019. *Access to Justice in European Union Law: A Legal Guide on Access to Justice in Environmental Matters*. London: ClientEarth.

Beyerlin, U., 2015. Aligning International Environmental Governance with the "Aarhus Principles" and Participatory Human Rights. In Grear, A. and Kotzé, J.K. (Editors), *Research Handbook on Human Rights and the Environment*. Northampton: Edward Elgar Publishing, 333–352.

Bonine, J., 2003. "Access to Justice in Cases Involving Public Participation in Decision-Making" and "The Public's Right to Enforce Environmental Law". In Sec, S. (Editor), *Handbook on Access to Justice under the Aarhus Convention*. Hungary: The Regional Environmental Center for Central and Eastern Europe, 23–37.

Boyd, D., 2012. *The Environmental Rights Revolution: A Global Study of Constitutions, Human Rights, and the Environment*, Law and Society Series. Canada: University of British Columbia Press.

Collins, L., 2015. The United Nations, Human Rights and the Environment. In Grear, A. and Kotzé, J.K. (Editors), *Research Handbook on Human Rights and the Environment*. Northampton: Edward Elgar Publishing, 219–244.

Craig, P., Hofmann, H., Schneider, J.P. and Ziller, J., 2017. *ReNEUAL Model Rules on EU Administrative Procedure*. Oxford: Oxford University Press.

Darpö, J., 2013. *Effective Justice? Synthesis Report of the Study on the Implementation of Articles 9.3 and 9.4 of the Aarhus Convention in the Member States of the European Union*. Brussels: European Commission.

García Ureta, A., 1996. Protección del ambiente y acceso a los tribunales: asunto T-585/93, Greenpeace International y otros v. Comisión. *Revista de Derecho Urbanístico y Medio Ambiente*, 150, 113–146.

García Ureta, A., 2005. El Convenio de Aarhus: derecho de participación y de acceso a la justicia. In Margariños Compareid (coord), *Derecho al conocimiento y acceso a la información en las políticas de medio ambiente*. Madrid: INAP, 17–102.

Hey, E., 2015. The Interaction between Human Rights and the Environment in the European "Aarhus Space". In Grear, A. and Kotzé, J.K. (Editors), *Research Handbook on Human Rights and the Environment*. Northampton: Edward Elgar Publishing, 353–376.

IMPEL, 2000. *Complaint Procedures and Access to Justice for Citizens and NGOs in the Field of the Environment within the European Union. Final report*. Belgium: IMPEL.

Jans, J.H., 2013. Judicial Dialogue and Judicial Competition and Global Environmental Law: A Case Study on the UNECE Convention on Access to Information, Public Participation in Decision-Making and Access to Justice in Environmental Matters. In Jans, J., Macrory, R. and Moreno Molina, Á.M. (Editors), *National Courts and EU Environmental Law*. Groningen: Europa Law Publishing, 145–168.

Jans, J.H. and Vedder, H.H.B., 2011. *European Environmental Law. After Lisbon*, 4th ed. Groningen: Europa Law Publishing.

Jendrośka, J., 2005. Aarhus Convention and Community Law: The Interplay. *Journal for European Environmental & Planning Law*, 2(1), 12–21.

Jendrośka, J., 2006. Public Information and Participation in EU Environmental Law. Origins, Milestones and Trends. In Macrory, R. (Editor), *Reflections on 30 Years of EU Environmental Law. A High Level of Protection?* Groningen: Europa Law Publishing, 63–86.

Juste Ruiz, J. and Castillo Daudí, M., 2014. *La protección del medio ambiente en el ámbito internacional y en la Unión Europea*. Valencia: Tirant Lo Blanch.

Krämer, L., 1997. *Focus on European Environmental Law*, 2n ed. London: Sweet & Maxwell.

Krämer, L., 2004. Acceso a la justicia ambiental en Europa. *IeZ: Ingurugiroa eta zuzenbidea/Ambiente y Derecho*, 2, 11–32.

Krämer, L., 2012. *EU Environmental Law*, 7th ed. London: Sweet & Maxwell.

Krämer, L., 2014. EU Enforcement of Environmental Laws: From Great Principles to Daily Practice – Improving Citizen Involvement. *Environmental Policy and Law*, 44, 247–256.

Krämer, L., 2018. Citizens Rights and Administrations' Duties in Environmental Matters: 20 Years of the Aarhus Convention. *Revista Catalana de Derecho Ambiental*, IX(1), 1–26.

Kravchenko, S. and Bonine, J.B., 2008. *Human Rights and the Environment: Cases, Law, and Policy*. Durham: Carolina Academic Press.

Lenaerts, K. and Gutiérrez-Fons, J.A., 2011. *The General System of EU Environmental Law Enforcement, Yearbook of European Law 2011*. London: Oxford University Press.

Martínez Jiménez, G., 2018. *Acceso a la justicia ambiental ante el Tribunal de Justicia de la Unión Europea. Aplicación del Convenio de Aarhus a las instituciones de la Unión*. Madrid: Centro de Estudios Políticos y Constitucionales.

Nagy, A.D., 2018. The Aarhus-Acquis in the EU. In Caranta, R. et al. (Editors), *The Making of a New European Legal Culture: The Aarhus Convention*. Groningen: Europa Law Publishing, 19–69.

Ortega Gómez, M., 2006. Legitimación de las asociaciones constituidas para promover los intereses colectivos de una categoría de justiciables ante el Tribunal de Justicia y el Tribunal de Primera Instancia. *Revista de Derecho Comunitario Europeo*, 25, 757–797.

Peñalver i Cabré, A., 2008. Novedades en el acceso a la justicia y la tutela administrativa en asuntos medioambientales. In Pigrau, A. (coord), *Comentario a la Legislación de acceso a la información, participación pública y acceso a la justicia en materia de medio ambiente* (*Ley 27/2006, de 18 de julio, por la que se regulan los derechos de acceso a la información ambiental, de participación pública y de acceso a la justicia en materia de medio ambiente*). Barcelona: Atelier, 349–403.

Peñalver i Cabré, A., 2016. *La defensa de los intereses colectivos en el contencioso-administrativo: legitimación y limitaciones económicas*. Navarra: Thomson-Reuters Aranzadi.

Peñalver i Cabré, A., 2019. El impacto del Derecho de la Unión Europea en el acceso al contencioso-administrativo español ambiental. In López Ramón, F. and Valero Torrijos, J. (coords), *20 años de la Ley de lo contencioso-administrativo. Actas del XIV Congreso de la Asociación Española de Profesores de Derecho Administrativo*. Madrid: INAP, 121–135.

Plaza Martín, C., 2005. *Derecho ambiental de la Unión Europea*. València: Tirant lo Blanch.

Poncelet, C., 2012. Access to Justice in Environmental Matters – Does the European Union Comply with its Obligations? *Journal of Environmental Law*, 24, 287–310.

Razquín Lizarraga, J.A. and Ruíz de Apodaca Espinosa, Á.M., 2007. *Información, Participación y Justicia en materia de medio ambiente (Comentario sistemático a la Ley 27/2006, de 18 de julio)*. Navarra: Thomson Aranzadi.

Rehbinder, E. and Stewart, R., 1985. Environmental Protection Policy. In Cappelletti, M. et al. (Editors), *Integration through Law: Europe and the American Federal Experiences*. Berlin: Walter de Gruyter.

Ruiz de Apodaca Espinosa, A.M., 2018. El acceso a la justicia ambiental a nivel comunitario y en España veinte años después del Convenio de Aarhus. *Revista Catalana de Derecho Ambiental*, IX(1), 1–53.

Salazar Ortuño, E., 2019. *El acceso a la justicia ambiental a partir del Convenio de Aarhus. Justicia ambiental de la transición ecológica*. Navarra: Thomson Reuters Aranzadi.

Sands, P. et al., 2018. *International Environmental Law*, 4th ed. Cambridge: Cambridge University Press.

Sec, S., 2003. Conclusions and Recommendations. In Sec, S. (Editor), *Handbook on Access to Justice under the Aarhus Convention*. Szentendre Hungary: The Regional Environmental Center for Central and Eastern Europe.

Sec, S. and Jendrośka, J., 2019. The Escazu Agreement and the Regional Approach to Rio Principle 10: Process, Innovation, and Shortcomings. *Journal of Environmental Law*, 31, 533–545.

Sommermann, K.P., 2017. Transformative Effects of the Aarhus Convention in Europe. *ZaöRV*, 77, 321–337.

United Nations Economic Commission for Europe (UNECE), 2014. *The Aarhus Convention: An Implementation Guide*, 2nd ed. New York and Geneva: United Nations.

United Nations Environment Programme, 2010. *Guidelines for the Development of National Legislation on Access to Information, Public Participation and Access to Justice in Environmental Matters*. Nairobi: United Nations.

Wates, J., 2005. The Aarhus Convention: A Driving Force for Environmental Democracy. *Journal for European and Planning Law*, 1, 2–11.

7 Environmental refugees

Reshaping the borders of migration in the EU

Susana Borràs-Pentinat

1. Introductory remarks

The so-called "migration crisis" is one of the most important challenges that face many EU member states today. People moving in a context of environmental degradation are part of such migration. The international response, including that of the European Union (EU), has been limited, and protection for people on the move due to environmental degradation remains inadequate, especially in a context of climate emergency in which millions of people are being forced to leave their homes because of environmental disruptions. Not only is there no recognition of the "environmental refugee", there is not a comprehensive multilateral framework of protection for those fleeing from increasing environmental degradation. Meanwhile, thousands of people have lost their lives at sea trying to reach the EU and nearly 90% of refugees and migrants have paid organized crime and human traffickers to cross borders.[1]

This chapter argues for the need to recognize and protect environmental refugees at the EU level. It also outlines an assessment of the limitations and loopholes that exist in the current EU regulatory framework in light of these challenges, and in the way the EU responds internally and internationally, seeking institutional mechanisms to strengthen migration governance within the framework of the global environmental crisis.

In order to examine these issues, the chapter will first address the root causes of the migrant crisis in the EU and its consequences. This will allow us, secondly, to focus on the legal and political aspects of environment-related migration and on the extent to which the current framework of the EU's asylum and migration policy offers adequate responses, internally and internationally, to environmental migration. Accordingly, the third part of the chapter will explore the current limitations of the EU to cope with environmental migration. These limitations contribute to the difficult situation in which people who help migrants, including victims of environmental degradation, are exposed to lawsuits, arrest, and violence. In many EU countries, new trends towards the criminalization of people who defend the human rights of migrants, refugees, and asylum seekers – "migrant human rights defenders" – endanger the fundamental values of the EU and aggravate the migration situation, resulting in a humanitarian emergency that requires urgent

EU intervention. Questions arise, however, as to whether the EU has the power to act and sufficient political will to do so. Finally, given that the current legal framework grants potential relief, the chapter will offer a series of proposals and specific arrangements that are needed by the EU to adapt its current migration policy and regulatory framework in order to protect and defend the human rights of people on the move due to environmental degradation.

2. People moving in the context of environmental change

Climate change intensifies the frequency and severity of disasters, thereby increasing the number of people displaced by extreme weather events.[2] Adverse climate impacts are already exacerbating patterns of human mobility, and will do so to an even greater degree in the future.[3]

Approximately 265 million people have been displaced by natural hazards since 2008.[4] The numbers have increased over the time: around 17 million people were internally displaced by disasters in 2018 alone. Generally, while the vast majority of climate migrants are displaced within their home countries, many people on the move are forced to cross international borders. The United Nations High Commissioner for Refugees (UNHCR) has estimated that at least some 30 million people were displaced in 2013 as a result of weather changes and natural disasters; and that there are currently around 50 million refugees in the world for whom the causes of their displacement can be connected in many cases to climatic alterations.[5]

Moreover, there is ample evidence that environment-related migration and displacement is already occurring and will increase in the future. Estimates that predict how many people will relocate because of climate change can vary from 25 million to one billion by mid-century, according to the International Organization for Migration (IOM).[6] In this respect, the United Nations Convention to Combat Desertification[7] (UNCCD) warns that by 2045, 135 million people may be displaced by desertification alone.[8] The Christian Aid (2007) group report also estimates that at least one billion people will be forced to leave their homes by 2050 in the face of a shortage of natural resources caused by climate change. The very disparity in the figures on potential environmental migrants is indicative of the complexity and uncertainty of the phenomenon of environmental migration.

IOM (2007) holds that several factors have an influence on migratory contexts and that it is sometimes difficult to separate the triggers of climate change from political, social, and economic aspects, among others. However, the environment and climate change are determining factors that play a notable and increasingly decisive role in human mobility. The growth in the frequency and intensity of sudden and gradual climate-related natural disasters implies a greater likelihood of humanitarian emergencies that will result in population displacements. The adverse consequences of warming, climate variability, and other effects of climate change on living conditions, public health, food security, and water availability can exacerbate pre-existing vulnerabilities and encourage migration. Rising sea levels can lead to the uninhabitability of coastal areas and low-lying islands. The scarcity of natural resources generates tensions and eventually conflicts, and, in turn, forced migration.

Despite the evidence that environmental changes, natural disasters, and degradation are drivers of migration, there is as of yet no internationally accepted definition of persons on the move due to environmental reasons. The lack of recognition and, therefore, of protection is partly due to the absence at the international level of the term "environmental refugee".[9] Indeed, the 1951 Geneva Convention Relating to the Status of Refugees[10] (1951 Refugee Convention) defines a "refugee" as a person who has crossed an international border "owing to well-founded fear of being persecuted for reasons of race, religion, nationality, membership of a particular social group or political opinion". However, to condition the protection of migrants solely on the existence of political causes is to simplify realities that are, in most cases, very complex.[11]

Undoubtedly, environmental migration is not only a controversial issue, but also one that requires debate now because it has ceased to be a likelihood and is now tangible. Indeed, there are people who have already had to emigrate because they have run out of resources and/or become landless because of environmental disasters. The 2017 Atlantic hurricane season demonstrates the extent of the risk of forced migration in the Caribbean; three of the season's major hurricanes, Harvey, Irma, and Maria, forced approximately three million people to move in a single month.[12]

However, people migrating for environmental reasons do not fall squarely within any one particular category provided by the existing international legal framework. Terms such as "environmental refugee" and "climate change refugee" have no legal basis in international refugee law. Indeed, there is a growing consensus among concerned agencies, including UNHCR, that their use is to be avoided.[13] This rejection revolves around the fact that UNHCR has no mandate to protect other types of migrants and that mixing groups in the same position would confuse and possibly undermine efforts to help and protect refugees. On several occasions, UNHCR has expressed its views on the issue of environmental refugees. For example, Ogata asserted that

> using the term environmental refugee to refer to all people forced to leave their homes because of environmental change loses the distinctive need of refugees for protection. It blurs the respective responsibilities of national governments towards their citizens and of the international community towards those who are without protection. It also impedes a meaningful consideration of solutions and action on behalf of the different groups. Therefore, UNHCR believes the term environmental refugee is a misnomer.[14]

In the same vein, IOM has put forward a broad working definition:

> Environmental migrants are persons or groups of persons who, predominantly for reasons of sudden or progressive change in the environment that adversely affects their lives or living conditions, are obliged to leave their

habitual homes, or choose to do so, either temporarily or permanently, and who move either within their country or abroad.[15]

The issue remains a matter of concern: the ill-defined legal situation of migrants due to environmental causes, together with a lack of responses, only increases their vulnerability, insofar as the solutions fail to address the causes either politically or legally. The main reason for this is the difficulty of isolating climate change and environmental deterioration from other variables that influence migration, such as economic ones.

As the Foresight Report (2011) notes, "the range and complexity of the interactions" among drivers "means that it will rarely be possible to distinguish individuals for whom environmental actors are the sole driver". Environmental degradation cannot be considered as an isolated factor of migration, mainly because there is a connection between socioeconomic, cultural, political, and social factors and the environment. Accordingly, the causes of migration are usually complex, include multiple factors, and depend on the characteristics of the environment and the particular circumstances of each affected person. A wide range of reasons, both natural and anthropogenic in origin, can produce environmental degradation and migration. Foremost in importance among the natural causes of environmental degradation are natural disasters, which include geophysical activities (earthquakes, volcanic eruptions, avalanches, and landslides), meteorological activities (tropical cyclones, tornadoes, hurricanes, and typhoons) and hydrological activities (floods). In general, however, economic factors play a larger role than environmental drivers in determining migration outcomes at both the individual and structural levels.

Although international processes like the Global Compact on Safe, Orderly and Regular Migration,[16] the Paris Agreement,[17] and the Sendai Framework for Disaster Risk Reduction[18] have emphasized the importance of environmentally induced migration, there is currently no comprehensive multilateral framework to guarantee protection to anyone outside the scope of international refugee law. Notwithstanding the fact these people on the move are victims and witnesses of environmental degradation, most cases are the result of activity by the richest countries in the Global North.

Regional solutions are widely seen as the most useful tool to provide a minimum of protection for environmental refugees, mainly because in most cases the phenomenon is internal or happens between bordering states. Regional instruments, such as those in Africa and America, have expanded the definition of refugee to persons fleeing "events seriously disturbing public order" (1969 OAU Convention; 1984 Cartagena Declaration). The EU's potentially crucial role and leadership on environmental and climate policies could be useful in closing this gap in protection for environmentally induced migration, providing a new opportunity for the EU to be a global player in tackling environmental migration through a new legal and institutional framework for the integration of environmentally displaced populations and environmental refugees.

Ultimately, the purpose of this chapter is to analyse how the debate on environmental migration is dealt with in the EU and to examine the extent to which environment-related migration will affect European countries and the EU from a legal and political perspective. In this respect, it is important to consider the EU's potential international and internal responses to the complex relationship between environmental degradation, including climate change, and people on the move. This will make it possible to analyse whether existing legal and policy frameworks are adequate to accommodate and cover different forms of environment-related migration effectively, and to determine whether new instruments and policies are needed.

In sum, the chapter will take a critical look at the EU's role and challenges in the face of this issue with the aim of contributing to a reconsideration of the concept of migration, treating it more as an experience of dignity than as an act of despair.

3. Environmental refugees: the forgotten migrants in the EU migration policy

The migrant crisis started in 2015 and 2016, when an estimated one million people arrived, mainly from Africa and the Middle East. Since 2014, the EU has replaced Mediterranean search-and-rescue missions with operations whose primary focus is security and border enforcement.

The influx of immigrants and asylum seekers to Europe has highlighted the need for more just and effective European asylum and migration policies. Moreover, the crisis has arisen not only because of push factors at origin, but also because of the fact that international law does not confer a general right of entry into a foreign country,[19] except for migrants who qualify for protection under refugee law or complementary measures.

Environmental refugees find themselves with no right to enter another state or be protected against forcible return. Their situation and their status depend upon the generosity of host countries. The environmental refugee may be afforded a right of entry when environmental drivers can be combined with well-established grounds for protection under the 1951 Refugee Convention. In fact, most migrants try desperately to prove political grounds for their exodus. Conversely, host countries treat suspected non-political grounds as a reason to force their return. In fact, fleeing a slow or sudden-onset disaster does not trigger any grounds for persecution due to race, religion, nationality, political opinion, or membership in a particular social group under the 1951 Refugee Convention. While such grounds do protect a small number of migrants, they also serve to exclude millions from protection. Nevertheless, environmental degradation also has an economic impact, which can involve processes of political persecution.

In recent years, many people have fled to Europe to escape conflict, terrorism, and persecution. Of the 333,355 asylum seekers who were granted protection status in the EU in 2018, more than a quarter came from war-torn Syria, followed by Afghanistan and Iraq in second and third place, respectively. In each of these countries, civilians face threats such as armed conflict, human rights violations, and persecution. Complex, interrelated environmental changes, such as droughts and floods, overexploitation of resources, and climate change are factors that contribute to cyclic mobility from rural areas to urban areas within the region's countries and across their borders. Some studies, led for example by Kelley[20] and Gleick,[21] have highlighted the relationship between, on one hand, the effects of climate change on livelihoods that depend on natural resources and food insecurity and, on the other hand, tensions, conflicts, and mobility. This can be seen, for instance, in the connections between drought, migration, and the Syrian conflict. Drought helped to displace Syria's farmers, contributing to the instability that triggered the country's civil war.[22]

Whatever the push factors may be, migrants are confronted at their destination by militarization, the securitization of borders, the construction of walls, the externalization of border controls, and other limitations that fail to consider their situations of humanitarian emergency.[23] Under these circumstances, only a few migrants will be protected as refugees, while the rest remain invisible as environmental migration. Moreover, on several occasions, major policy decisions such as the "renationalization" of migration policy and the refusal to take legal responsibility have led the EU to jeopardize its own treaty-based framework, which aims to ensure the democratic rule of law and fundamental rights. One example comes with the EU-Turkey Declaration on the Migration Crisis and the subsequent Order of the General Court of the EU, which found that it lacked jurisdiction to hear and determine the actions brought by three asylum seekers against the EU-Turkey Statement.[24] A second example concerns how some EU member states have failed to meet their relocation quotas according to Council Decision (EU) 2015/1523 of 14 September, establishing provisional measures in the area of international protection for the benefit of Italy and Greece,[25] and Council Decision (EU) 2015/1601 of 22 September, establishing provisional measures in the field of international protection for the benefit of Italy and Greece.[26] In the case of Spain, the Judgment of the Supreme Court (section 5 of the Administrative Litigation Chamber) of 9 July 2018 found that the Spanish state must fulfil its quotas.[27] In sum, these examples corroborate that national interests prevail over humanitarian ones. Rather than working to tackle the causes of the refugee problem, the EU has made a political decision to prevent refugees from entering EU countries, which have been stopping migrant boats and stepping up fines for human trafficking. And, finally, those who survived and managed to enter by human quotas are ultimately rejected by the states.

As happens internationally, there is currently no comprehensive approach to the phenomenon at the EU level, thereby contributing to the threat posed to the living conditions and the very lives of thousands of persons. Nor are there are appropriate policy responses or a well-structured regulatory framework to address environment-related migration.

The EU policy on migration, including environmental migration, has an internal and an external dimension. Both dimensions are strongly influenced by the objective of limiting or managing immigration and refugee flows to EU territory. The internal dimension relates to migration into the EU. The external dimension refers to the inclusion of migration and asylum policies in external relations with third countries. The latter aims at preventing immigration by addressing the root causes of migration in third countries and by transferring migration control instruments – such as border control or measures to combat illegal migration, smuggling, and trafficking – outside of the EU.

The justification for this approach is that environment-related migration will primarily occur "in the developing world, either with migrants moving internally or to countries in the same region".[28] However, the EU fails to acknowledge, on one hand, that there is no data available on whether and how many people are moving towards Europe for, *inter alia*, environmental reasons. It also ignores, on the other hand, that people who would like to move to the EU in this context would not be able, in most cases, to do so legally – especially in the case of unskilled and poor people.

Article 67(2) of the Treaty on the Functioning of the European Union (TFEU), which is included in the general provisions of the title relating to the area of freedom, security, and justice, provides in general that the EU "will develop a common policy of asylum, immigration and control of external borders that is based on solidarity between Member States and is equitable with respect to third-country nationals". Thus, in the terms of the precept itself, a "common policy" is to be set up to cover three different subjects: border control, asylum, and immigration. It should be stressed that we are not looking at an area subject to the exclusive competence of the EU.[29] Rather, the use of the term "common policy" points clearly to a forward-looking willingness to embrace expansion, coordination, and continuity that may prove important in the future. Expanding this policy to environment-related migration, however, represents a huge challenge.

In this sense, although the treaties can provide a sufficiently broad mandate for a review of asylum and immigration policy to regulate the status of "environmental migrants", there is neither an EU instrument nor a coherent policy that mentions them. However, articles 77 to 80 of the TFEU provide a general mandate for the development of common policies in all major areas of immigration and asylum policy. Article 80 of the TFEU puts the principle of solidarity and fair sharing of responsibility at the very heart of the entire Union system, providing a legal basis for the implementation of these principles in the Union's policies on asylum, migration, and control of borders.

The Council Directive 2004/83/EC of 29 April 2004, which addresses the minimum standards for the qualification and status of third-country nationals or stateless persons as refugees or as persons who otherwise need international protection and the content of the protection granted (the "Qualification Directive"),[30] sets out who qualifies for international protection and who does not. It is unlikely that people displaced by the effects of climate change would qualify for protection under the directive.

The Council Directive 2001/55/EC of 20 July 2001, which covers both the minimum standards for giving temporary protection in the event of a mass influx of displaced persons and measures to promote a balance of efforts between member states in receiving such persons and bearing the consequences thereof (Temporary

Protection Directive – TPD),[31] establishes minimum standards for giving temporary protection to environmental refugees under certain conditions. However, this protection was created for the exceptional circumstance of a mass influx of displaced persons, and is thus not available to individual situations. It is only to be applied for a large group of people fleeing from an armed conflict, or for people who are at serious risk of, or are victims of, systematic or generalized violations of human rights. Even if this system were to be applied, people displaced due to permanent or long-lasting environmental damage would still not receive any protection.

While such protection is entirely at the discretion of each EU member state, environment-related forms of migration are also a topic of discussion at the EU level. Accordingly, the issue will be analysed in the next section.

The EU's approach to environment-related migration was initially driven largely by security considerations; however, it has lately been complemented by a development approach. The former refers to the perception of immigration primarily as a threat to national security. The latter, by contrast, highlights migration as a positive contribution to the development of a region. While the development approach emphasizes the positive aspects of migration, it has also been criticized for seeking to pursue the same objectives as the security approach, i.e., to strengthen immigration control and to decrease migration towards European countries.

In this respect, the Commission, through DG DEVCO (rather than DG HOME), has published a staff working paper on the EU Adaptation Strategy, which makes an assumption that no one will migrate to Europe in such circumstances. However, given that four-fifths of the worldwide refugee population stay within their regions of origin, it is clear that the combination of population growth, an increase in the quantity and intensity of environmental hazards, and insufficient adaptive capacity will put ever greater pressure on their regions of origin. Environmental migration, therefore, is seen as adding a new dimension to development policy. While this attests to the low political priority of the matter, it also offers real potential for a holistic treatment of the issue within development strategies, which would accord better with the reality of the phenomenon.

Both dimensions are problematic, however, because they feed into stereotypes of a mass influx of poor, mostly unskilled migrants or, paradoxically, because they deny responsibility for any activity in the area of EU immigration and asylum policy by regarding it as primarily a "development" problem, based on the claim that most migration occurs within the Global South.[32]

As noted earlier, there is also a growing trend towards the criminalization of humanitarian assistance to immigrants and asylum seekers. The lack of appropriate recognition for environmentally induced migration contributes to a lack of protection, given the irregularity of the situation. Thus, ultimately, it remains in the hands of NGOs and individuals to save the lives and guarantee the rights of migrants. However, the existing restrictions on their search and rescue activities, together with an increasing criminalization of their efforts, aggravates the situation of forgotten environmental migrants. Overall, this situation is exemplified by an intensification of the EU's restrictive approach to migration policy since late 2014: the joint EU hostility towards migrants and refugees can lead to arrest, legal troubles, or harassment that is far removed from understanding the root causes of migrations. This approach poses a clear constraint on the need to extend protection to people fleeing from environmental disruptions.

4. Towards the recognition of environmental refugees at the EU level

On the whole, decision-makers have responded reactively to concerns about migration linked to environmental change, generally avoiding the issue and, wherever possible, lowering its priority within international discussions. Where governments have been forced into a policy decision (such as occasions when a natural disaster has led directly to migration), policy makers have framed their responses as "one-off efforts" to deal with natural disasters and have carefully avoided setting precedents, while also emphasizing the goal of reducing the supposed threat of mass movement to their countries.

As in the case of the United Nations (UN) system, the EU has yet to develop a framework or instrument to recognize and protect environment-related migration. Nevertheless, an increased interest in the issue can be observed in recent years whereby the EU's political discourse has been evolving around the potential threat of the phenomenon to its own internal security. Precisely for this reason, European decision-makers should start considering migration linked to climate change and other environmental changes.

Indeed, over the past decade, EU institutions have commissioned and published an increasing number of studies and policy papers on the subject.[33] All of this progress and debate can help to clarify the EU's position, both internally and externally, towards environment-related migration. Nearly every EU institution started to engage with environment-related migration and displacement in the late 1990s by gathering information, commissioning research projects, and organizing events in order to discuss the topic with different stakeholders.

The initial attempt of the European Parliament

The European Parliament was the first EU institution to do some groundwork in this regard.[34] Two main limitations can be observed in the process: first, only Green Members of the European Parliament (MEPs) promoted a discursive framing of the issue; and second, the initiatives focused only on the phenomenon of "climate refugees". The lack of a broad political will and an incomplete picture of the issues, however, frustrated these efforts, which are described in the following.

In the European Parliament's 1999 Resolution on "The Environment, Security and Foreign Policy", the term "climate refugees" was included as a challenge for EU immigration and justice policies, development assistance, and spending on humanitarian aid as well as heightened security problems for the EU in the form of regional instability in other parts of the world.[35]

From 2001 onwards, several Green MEPs tried unsuccessfully to insert a reference to environmental refugees into a European Parliament report on the common European asylum policy. After several written declarations asking to ascribe community status to ecological refugees and to put the principle of ecological interference into community prerogatives, the European Parliament did not adopt any formal resolution.[36]

In April 2007, the European Parliament set up a Temporary Committee on Climate Change (CLIM) in order to formulate proposals on the EU's future integrated policy on climate change and to coordinate the Parliament's position in negotiations on the international framework for climate policy after 2012.[37] A vote on the final report took place in plenary session before the end of CLIM's mandate in February 2009.

The Greens/European Free Alliance (EFA) Group held a seminar on climate refugees in June 2008 in order to once again focus the attention of EU institutions on the need to recognize climate migrations. The outcome was a final Declaration, which ended with conclusions that called upon European institutions to develop a European strategy on climate-forced migration in order to "organize legal protection for the victims of climate disruptions and of possible displaced persons (current or future) who do not benefit today from any recognition".[38] Additionally, there was a call to launch a debate within the UN on the status of climate migrants and on a protocol to the UNFCCC on climate-forced migration.

A study called "Climate Refugees: Legal and Policy Responses to Environmentally Induced Migration",[39] commissioned by the European Parliament (Committee on Civil Liberties, Justice and Home Affairs), explored protection gaps and possible policy responses, recommendations, and proposals for concrete actions to recognize climate refugees formally within the current EU legal and policy framework.[40]

In 2013, the Greens/EFA adopted a position paper entitled "Climate Change, Refugees and Migration",[41] which identified a number of EU policy instruments that could potentially be extended to include climate-induced migration and grant subsidiary protection status.

The European Parliament adopted the Resolution of 12 April 2016 on the situation in the Mediterranean and the need for a holistic EU approach to migration.[42] However, it was not until the 2017 Resolution, called "Addressing refugee and migrant movements: the role of EU external action",[43] that the European Parliament stressed that "EU development cooperation should continue to address and effectively tackle the root causes of forced displacement and migration", treating climate change as one of these root causes, in line with Goal 16 of the 2030 Agenda for Sustainable Development. This Resolution also called upon member states to take up their global responsibility in the face of climate change by implementing the Paris Agreement and adopting "a leading role in recognizing the impact of climate change on mass displacement, as the scale and frequency of displacements are likely to increase". Consequently, the Resolution demanded that the EU and its member states put more financial resources at the disposal of the countries most affected by climate change, and called for a special international protection status for those migrants resulting from climate events.

In its 2018 Resolution "On Progress on UN Global Compacts for Safe, Orderly and Regular Migration and on Refugees", the European Parliament called upon all UN members, but more specifically the EU countries, to focus particularly on the drivers of irregular migration and forced displacement, including climate change and natural disasters.[44]

The Greens/EFA Group in the European Parliament launched another initiative in June 2019, adopting a position paper called "Climate Change, Refugees and Migration".[45]

Unfortunately, the failed efforts described here remain as an anecdotal contribution and the political will to tackle the issue is still broadly lacking.

The cautious approach of the European Commission and the EU high representative

The approach of the European Commission to environment-related migration can be characterized as cautious and reserved, especially when it comes to migration and asylum policies and frameworks. The European Commission has been quite reluctant to take a leadership role at the international level or towards EU member states. To date, the suggested or adopted policies have aimed instead at promoting existing initiatives or reframing existing measures (e.g., development policies) and treating them as relevant for environment-related migration as well.[46]

From January 2007, the European Commission took hold of the matter through the financing of the EACH-FOR ("Environmental Change and Forced Migration Scenarios") project within the framework of FP6 (Priority 8.1 Policy-oriented research).

In March 2008, the high representative for the Common Foreign and Security Policy and the European Commission delivered a paper to the Council on "Climate Change and International Security", which asserted that climate change might cross over into other areas such as migration.[47] While no action was taken to move the policy agenda forward, the paper did describe environmentally induced migration as a "threat" that might enhance the potential for conflict in transit and destination countries. The paper also pointed out that Europe "must expect substantially increased migratory pressure" and recommended an enhancement of EU capabilities by building up knowledge and systems of monitoring and early warning. In addition, it emphasized the role of the multilateral leadership of the EU to promote global climate security by considering "environmentally-triggered additional migratory stress in the further development of a comprehensive European migration policy, in liaison with all relevant international bodies".

In January 2008, IOM and the United Nations Environment Programme (UNEP) arranged a common meeting called "Migration and Environment" in order to increase the European Commission's awareness and sensitivity to these issues. Several representatives of different directorates-general attended the discussion and showed a high level of interest.

Shortly afterwards, the European Commission adopted a Communication on "A Common Immigration Policy for Europe: Principles, Actions and Tools"[48] and a Policy Plan called "Asylum – an integrated approach to protection across the EU".[49] The Communication proposed ten common principles on which to build a common immigration policy. They were grouped under the three main strands of EU policies: prosperity, solidarity, and security. In parallel, but in a separate document, the Policy Plan on Asylum provided options for shaping the second phase of the Common European Asylum System (CEAS). The Policy Plan sought to improve the definition of protection standards at the EU level by amending, in the course of 2009, the currently existing legal instruments such as the directive on requirements for qualification as refugees or persons needing international protection. While the Policy Plan made no reference to environmental refugees or environmentally induced migrations, there was certainly a window of opportunity. An important trend identified in the Policy Plan is that

> an ever-growing percentage of applicants are granted subsidiary protection or other kinds of protection status based on national law, rather than refugee status according to the Geneva Convention. This is probably due to the fact that an increasing share of today's conflicts and persecutions are not covered by the Convention. It will therefore be important during the second phase of the CEAS to pay particular attention to subsidiary and other forms of protection.[50]

One of the overarching objectives of the CEAS is "to ensure access to those in need of protection". This undertaking might mark the start of the Commission's work on the issue of environmental refugees. Ultimately, the proposals endorsed by the European Council on 15 October 2008, initially led to the adoption of a European Pact on Immigration in view of the new five-year Programme in the Justice, Freedom and Security area, which was adopted in the second half of 2009.[51]

Similarly, the publication of a White Paper on adaptation to climate change in the autumn of 2008 could mark the beginning of the process. The White Paper was a follow-up to the European Commission Green Paper of 29 June 2007 on adaptation to climate change in Europe.[52]

The Action Plan for the implementation of the Stockholm Programme (2010–15) contained a point asking the European Commission to issue a Communication with specific reference to international climate-induced migration, including the possible effects on immigration in the European Union. However, the final Communication, which was called "Global Approach on Migration and Mobility", paid little attention to the matter.[53]

The European Commission's Staff Working Document (CSWD) on "Climate change, environmental degradation and migration" (accompanying the aforementioned EU strategy on adaptation to climate change), which was published in April 2013, shifted the focus back to the Global South, stating that migratory flows to Europe due to climate stress were unlikely.[54] Consequently, the paper focused on recommendations for EU policies with an external perspective such as development, foreign policy, and humanitarian aid. After pointing out that most migratory movements would happen internally, the document concluded that there was no need for a "refugee-type protection" in regard to climate-induced migration.

Two years later, in 2015, the European Commission's Science for Environmental Policy series published "Migration in Response to Environmental Change". This document not only gathered all research findings, but it also acknowledged that migration patterns to the EU were affected by climate change, including EU citizens who might be moving among member states due to climate and environmental events. This latter consideration went beyond any consideration of "climate refugees" under current international law, since such situations of EU internal migration are mainly regulated by provisions on the free movement of persons, according to EU law, and to some extent by provisions set out in the EU Charter of Fundamental Rights. Therefore, the main challenge continued to be people arriving in the EU from outside.

The "European Agenda on Migration" adopted by the European Commission in 2015 identified climate change as a factor that directly and immediately fed into irregular migration and forced displacement, and it called for the "prevention and mitigation of these threats".

The president of the European Commission addressed the topic of climate-induced human mobility most notably in his State of the Union address in 2015 in the context of security concerns over climate refugees and climate migration. The president recognized that

> [i]n some parts of the world, climate change is changing the sources of conflict – the control over a dam or a lake can be more strategic than an oil refinery. Climate change is even one of the root causes of a new migration phenomenon. Climate refugees will become a new challenge – if we do not act swiftly.

After the developments of 2015 and subsequent years, migration and refugee issues have been addressed in the context of integration and efforts to decrease the numbers of new people arriving in EU territory.[55] Since 2016 and 2017, the topic of climate- and environment-related migration has become less prominent, though with some exceptions.

In 2017, for example, the European Commission's European Political Strategy Centre published "10 Trends Shaping Migration". This document recognized climate change not only as one of the ten trends in question, but also as the one most likely to dwarf all other drivers of migration, noting that it was already the case that "[m]any more people [are] internally displaced by 'natural' disasters than by violent conflicts and civil wars". In addition, the paper called for additional research on appropriate responses, finding that the current global humanitarian system was underfunded and ill-equipped to cope with future developments.

In his 2018 State of the Union, the president referred to climate change as a threat to the EU's security and affirmed the EU's global responsibility to tackle climate change, noting the summer droughts of 2018 in Europe.

The voluntary (im)prudence of the European Council

The connection between climate change, migration, and development prompted the European Commission to carry out "an analysis of the effects of climate change on international migration, including its potential effects on immigration to the Union". This served as a starting point for the development of the CSWD, published in April 2013. During the preparation of the CSWD, the European Commission held expert consultations and roundtables, such as the consultation of experts on "Climate Change and Migration" in May 2011. The CSWD was the culmination of the European Commission's efforts to launch a discussion on the interlinkages between migration, environmental degradation, and climate change, and to provide an overview of the research and data currently available on the subject. The CSWD's aim was to present a compilation "of the many initiatives of relevance for the topic which are already being taken by the EU in various policy fields" and to analyse "on-going debates on policy responses at EU and international level".

On 19 July 2013, the Council of the European Union released the Conclusions on the 2013 UN High-Level Dialogue on Migration and Development.[56] In relation to broadening the development-migration nexus, these Conclusions found that "climate and environmental degradation are already exerting an increasing influence on migration and mobility" and therefore that the interlinkages between climate change, environmental degradation, and migration should be further explored and addressed as appropriate, in particular in the context of development cooperation, foreign policy, and humanitarian assistance. The document called upon the EU and its member states to urgently take steps to deepen knowledge and further develop policy in this field.[57]

At the Council of the European Union in March 2008, Mr. Javier Solana predicted that "there will be millions of environmental migrants, with climate change as one of the major drivers of this phenomenon".[58] His address linked the matter to security considerations. Accordingly, he invited the Council to formulate recommendations before December 2008.

The presidency's Conclusions from the European Council meeting in Brussels on 19 and 20 June 2008 highlighted the following points:

> The EU is determined to provide an effective collective response to the new challenges to development posed in particular by climate change and high food prices. With regard to climate change, the EU is determined, where relevant, to help developing countries, particularly poor developing countries most vulnerable to climate change, to move towards sustainable economic growth and to adapt to climate change, in line with the agreement reached in Bali to launch negotiations aimed at securing a global agreement on climate

change in Copenhagen in 2009. It will work for the effective implementation of the 2007 "Global Climate Change Alliance" and will explore ways to mobilize new financial resources to tackle climate change and combat its negative impact. In this context, the EU will work, inter alia, on the basis of the Commission proposal for a global financing mechanism.[59]

Further additional political steps were set out in the Stockholm Programme,[60] and adopted by the European Council in December 2009.[61] The Programme provided a framework for EU action on the issues of citizenship, justice, security, asylum, immigration, and visa policy for the period 2010 to 2014. In particular, it gave greater attention to climate change as a driver of security-relevant migratory flows, stating specifically that

> [t]he connection between climate change, migration and development needs to be further explored, and the European Council therefore invites the Commission to present an analysis of the effects of climate change on international migration, including its potential effects on immigration to the European Union.[62]

5. Mind the gap: defending and protecting environmental refugees at the EU level

The Global Compact on Safe, Orderly and Regular Migration identified climate change as a driver of migration and suggested that countries work together to start planning for people who move due to natural disasters and climate change. To implement this objective, the EU as a whole and its member states should consider further steps to ensure changes in the "existing legal instruments" as ways to explore protecting and accommodating the category of environmental migration.

The defence and protection of environment-related migration requires the translation of research-based evidence and recommendations into practical policy. The EU could act as a leading political force and should begin to develop a holistic policy approach that addresses climate change and migration. In this sense, several options may be pursued to protect environmental migration, from complementary protection through resettlement actions to other alternatives under the EU's migration policy.[63]

The most feasible solution to provide protection for environmental migration would likely be the enlargement of subsidiary protection[64] done by modifying the directives related to migration by inserting a reference to environmental refugees. In this respect, the Qualification Directive could be amended by broadening the definition of "serious harm" to cover harm resulting from rapid-onset natural disasters. Indeed, the Directive already grants subsidiary protection to a:

> [T]hird country national or a stateless person who does not qualify as a refugee but in respect of whom substantial grounds have been shown for believing that the person concerned, if returned to his or her country of origin,

or in the case of a stateless person, to his or her country of former habitual residence, would face a real risk of suffering serious harm . . . and is unable, or, owing to such risk, unwilling to avail himself or herself of the protection of that country.

Under this definition, it would be possible to include environmental migration as well.

In addition, the Temporary Protection Directive could be modified to provide some protection not only to large groups of people, but also to other smaller groups of people.

As for the external dimension of the EU's migration policy, the European Commission held a consultation in 2011 on the relationship between climate change and migration. This consultation resulted in a review of the EU Global Approach to Migration, which looked at both the existing EU legal framework and the concept of "responsibility to protect" developed by the International Commission on Intervention and State Sovereignty. It also looked into the possible expansion of labour exchange agreements as a way to respond to slow-onset events, together with the possibility that the EU Development Corporation could offer more adaptation funding focused on preventing migration and treating migration as adaptation.[65]

Even as this process is expected to happen, however, very few member states have implemented systems of temporary protection or introduced protection for environmentally displaced people in their national legislation. For instance, Cyprus, Finland, Italy, and Sweden have legislation that explicitly offers some protection for people displaced by natural disasters, while other member states like Belgium, Bulgaria, Spain, Ireland, Lithuania, Latvia, Malta, and Slovakia have legislation that could possibly be interpreted as providing protection.

Part of this complementary protection could also come from the development of EU Regional Protection Programmes (RPPs) to reinforce the external asylum dimension as a potential solution for specific regions of the world.[66] RPPs aim to focus funding on relief, rehabilitation, and regional development. They furnish support for such measures as the provision of registration and protection services, cooperation on legal migration matters, and agreements on returns of failed asylum seekers or illegal immigrants to countries outside the EU.

Resettlement could offer another option to expand protection within the context of a Common European Asylum System. According to the 2009 European Commission Communication on the "Establishment of a Joint EU Resettlement Programme", member states can claim "compensation" from the EU under the European Refugee Fund for every person they resettle, with priority given to several categories of especially vulnerable people.[67] The current inflexibility of such categories, however, prevents the fund from responding to newly arising needs, such as people on the move due to the effects of climate change, although this difficulty could be addressed through a regular review and updating of priorities.[68] If environmental migration were included, for instance, it would provide a financial incentive for countries to offer resettlement and avoid the violation of human rights.

In addition, the European Declaration of Human Rights and the Charter of Fundamental Rights of the European Union offer viable foundations for the protection of environmental migrants. In the future, it might also be possible to invoke aspects of migration and refugee legislation, such as the right to family reunion and non-refoulement, in order to establish an EU-wide policy governing the expansion of existing visas to people who may be at risk of environmental hazards if they return. However, neither of these options specifically protects people who have been displaced by environmental degradation or natural disasters in the first place.[69]

In the latter case, some states have attempted to identify solutions or at least have given consideration to the matter. The principle of subsidiarity allows states to apply additional protection standards, consistent in nature with the EU's shared competence on borders, asylum, and immigration, and with the need to adapt migration policy to different social, economic, and environmental realities. According to article 5.3 of the TEU,

> in areas which do not fall within its exclusive competence, the EU shall act only if and in so far as the objectives of the proposed action cannot be sufficiently achieved by the Member States, either at central level or at regional and local level, but can rather, by reason of the scale or effects of the proposed action, be better achieved at Union level.[70]

Only a few member states have set up a mechanism of protection for persons who flee an environmental disaster. For example, Sweden[71] and Finland[72] provide for the recognition of asylum or humanitarian protection for those who cannot return to their country because of a natural disaster. Pursuant to the Swedish Aliens Act (Chapter 3 Section 3) and the Finnish Aliens Act (Chapter 6 Section 88), a person who has left his or her native country because of an environmental disaster may also qualify for asylum and therefore as a person with a "need for protection". The Belgian Senate's Committee on Foreign Relations and Defence adopted a resolution in 2006 that called for the promotion of the recognition of the status of environmental refugees in the relevant international conventions. In addition, the Italian Home Office decided in 2008 to suspend repatriation measures against Bangladeshi citizens residing in the country illegally in the wake of the crisis following Cyclone Sidr. Unlike Sweden and Finland, however, Italy has not granted any form of special protection or residence permit. Indeed, most member states have been unwilling to accept this kind of "environmental/climate asylum" and fail to see the point.

As a potential further step, the UN Human Rights Committee[73] has recently reached a decision whereby governments must take into account human rights violations caused by the climate crisis when considering the deportation of asylum seekers. In order words, a state will be in breach of its human rights obligations, in particular articles 6.1 and 7 of the International Convention on Civil and Political Rights, if it returns someone to a country where, due to the climate crisis, their life is at risk or in danger of cruel, inhuman, or degrading treatment.

6. Final remarks

Europe faces a complex paradox. On one hand, it has become one of the world's most important migration destinations in recent years against a backdrop of sectoral labour shortages and an aging workforce. On the other hand, migration continues to be prioritized in terms of security, deterrence, and the fight against illegal immigration.

A coordinated EU response has become especially evident in the present migration and refugee crisis. Within this context, environment-related migration presents new challenges for policy makers at both international and national levels. The analysis in this chapter suggests that climate change, or even environment-related migration, when compared to other issues, is a rather low-priority item on the EU policy agenda.

Through the Temporary Protection Directive, the current legal framework could grant a potential relief, but this protection is limited to cases of mass influx migration due to rapid-onset natural disasters, leaving aside other realities of environment-related migration. In addition, the lack of the activation of this Directive impedes a coordinate response at a national level to cope with the migration crisis.

Consequently, there is a need for a coherent and binding solution, including appropriate definitions, rules, institutions, and funding at the EU and international levels to meet the challenge. The process thus far, however, suggests that only incremental and less controversial policy changes will emerge, for example, in the fields of development cooperation and humanitarian aid, rather than any development of new policies or a review and opening up of legal instruments in the area of migration and asylum policy.

Notes

1 Europol, 2016. Trafficking in human beings in the EU. The Hague: Europol Public Information.
2 Rigaud et al., 2018, pp. 23, 107; Black et al., 2013.
3 IPCC, 1998; Boano et al., 2008, p. 12.
4 The Internal Displacement Monitoring Center, IDMC, 2019, p. 37.
5 Norwegian Refugee Council, 2014, p. 8; UNHCR, 2017, p. 12.
6 IOM, 2007, pp. 11–12; Stern, 2006, p. 3.
7 UNTS, vol. 1954, p. 3.
8 IOM, UNCCD and Vigil, 2019, p. 13; Global Humanitarian Forum (GHF), 2009, p. 35.
9 Tripathi, 2017, p. 23.
10 UNTS, vol. 189, p. 137.
11 Peers, 2011, pp. 417–427; Pascouau, 2013; Crisp, 2003.
12 GRID, 2018, p. 42.
13 UNHCR organized an expert roundtable on climate change and displacement, which was held in Bellagio, Italy, from 22 to 25 February 2011, with the support of the

Rockefeller Foundation. According to the "Summary of Deliberations on Climate Change and Displacement", one of the messages was: "The terms of 'climate refugee' and 'environmental refugee' should be avoided as they are inaccurate and misleading", available from: www.unhcr.org/4da2b5e19.pdf [accessed 18 November 2019].

14 See Statement by Mrs. Sadako Ogata, United Nations High Commissioner for Refugees, at the Swiss Peace Foundation, Geneva, 30 October 1992, available from: www.unhcr.org/admin/hcspeeches/3ae68fad20/statement-mrs-sadako-ogata-united-nations-high-commissioner-refugees-swiss.html [accessed 18 November 2019].

15 IOM, 2011, p. 33.

16 UN General Assembly, Global Compact for Safe, Orderly and Regular Migration, Marrakesh, Morocco, 10 and 11 December 2018, A/CONF.231/3, 30 July 2018.

17 U.N. Doc. FCCC/CP/2015/10/Add, 1 (29 January 2016).

18 A/RES/69/283.

19 Guy Goodwin-Gill, 1978.

20 Kelley et al., 2015.

21 Gleick, 2014, pp. 331–340.

22 Kelley et al., 2015, pp. 3241–3246.

23 De Haas, 2018.

24 Orders of the General Court in Cases T-192/16, T-193/16 and T-257/16 NF, NG and NM v European Council, 28 February 2017, ECLI:EU:T:2017:128, ECLI:EU:T:2017:129 and ECLI:EU:T:2017:130.

25 OJ No. L 239, 15.9.2015.

26 OJ No. L 248, 24.9.2015.

27 STS 2546/2018 – ECLI: ES:TS:2018:2546.

28 European Commission Staff Working Document (CSWD): Climate change, environmental degradation, and migration, SWD(2013)138 final, 16.4.201.

29 Pascouau, 2013.

30 OJ No. L 304, 30.9.2004.

31 Council Directive 2001/55/EC of 20 July 2001 on minimum standards for giving temporary protection in the event of a mass influx of displaced persons and on measures promoting a balance of efforts between Member States in receiving such persons and bearing the consequences thereof (Temporary Protection Directive, OJ No. L 212, 7.8.2001. See also Directive 2011/95/EU of 13 December 2011 on standards for the qualification of third-country nationals or stateless persons as beneficiaries of international protection, for a uniform status for refugees or for persons eligible for subsidiary protection, and for the content of the protection granted (recast), OJ No. L 337.

32 Black et al., 2011, pp. 447–449.

33 Geddes and Sommerville, 2013.

34 Sgro, 2008.

35 European Parliament Resolution on the environment, security and foreign policy, A4–0005/99, 28.1.1999.

36 See in this regard European Parliament, Written declaration on the community status of ecological refugee, by Marie Anne Isler Beguin and Jean Lambert, 9 February 2004, DC\523175FR.doc EP 342.103 and European Parliament, Written declaration on the principle of ecological interference, by Marie Anne Isler Beguin, 24 March 2004, DC\530228FR.doc EP 344.467.

37 Sgro, 2008.

38 Declaration on Climate Migrations 2008, p. 4.

39 Kraler et al., 2011.

40 Petrillo, 2014.

41 The Greens/EFA in the European Parliament, Position paper climate change, refugees and migration, adopted in May 2013, drafted by Hélène Flautre, Jean Lambert, Ska Keller and Barbara Lochbihler, available from: https://europeangreens.eu/sites/

europeangreens.eu/files/news/files/Greens%20EFA%20-%20Position%20Paper%20-%20 Climate%20Change%20Refugees%20and%20Migration.pdf [accessed 18 November 2019]. Before this document, the European Parliament adopted in 2012 a Report, through its Committee on Women's Rights and Gender Equality, on Women and climate change, A7–0049/2012.

42 European Parliament, Resolution of 12 April 2016 on the situation in the Mediterranean and the need for a holistic EU approach to migration (2015/2095(INI)).

43 European Parliament, Position Paper of 5 April 2017 on addressing refugee and migrant movements: the role of EU External Action (2015/2342(INI)).

44 Available from: www.europarl.europa.eu/doceo/document/TA-8-2018-0118_EN.html? redirect [accessed 10 January 2020].

45 See the Working Paper Greens/EFA Group in the European Parliament (2019). Climate Refugees and Climate Migration, available from: https://gef.eu/wp-content/ uploads/2019/06/Climate-paper_FINAL.pdf [accessed 18 November 2019].

46 Mayrhofer and Ammer, 2014.

47 European Commission and the High Representative, Climate Change and International Security, Joint Paper to the European Council, 3 March 2008, 7249/08.

48 European Commission Communication: A Common immigration policy for Europe: principles, actions and tools, COM(2008)359 final, 17.6. 2008.

49 European Commission Communication: Policy Plan Asylum – an integrated approach to protection across the EU, 17.6.2008.

50 Policy Plan 2008:3.

51 Vlassopoulos, 2012.

52 European Commission Green Paper, Adaptation to climate change in Europe – options for EU actions, COM(2007)354 final, 29.6.2007. In this regard see Geddes and Sommerville, 2013.

53 European Commission Communication: The Global Approach to Migration and Mobility, COM(2011)743 final, 18.11.2011.

54 European Commission Staff Working Document (CSWD): Climate change, environmental degradation, and migration, SWD(2013)138 final, 16.4.2013.

55 De Haas, 2018.

56 Conclusions of the Council and of the Representatives of Governments of the Member States meeting within the Council on the 2013 UN High-Level Dialogue on Migration and Development and on broadening the development-migration nexus. 12415/13, 19.7.2013.

57 Vlassopoulos, 2012.

58 Commission and the Secretary-General/High Representative 2008, p. 5.

59 European Council Presidency Conclusions 2008, p. 18.

60 Council of the European Union, Draft Multi-annual programme for an area of Freedom, Security and Justice serving the citizen (The Stockholm Programme), 14449/09, JAI 679, 16.10.2009. Available from: https://register.consilium.europa.eu/doc/srv?l= EN&f=ST%2014449%202009%20INIT [accessed 10 January 2020].

61 European Council: The Stockholm Programme – An open and secure Europe serving and protecting the citizens, 16484/1/09 REV 1 JAI 866 + ADD 12009. Available from: https://ec.europa.eu/anti-trafficking/sites/antitrafficking/files/the_stockholm_pro gramme_-_an_open_and_secure_europe_en_0.pdf [accessed 10 January 2020].

62 *Ibid.*

63 Ammer et al., 2014.

64 Magniny, 1999.

65 Black et al., 2011; Boswell, 2003.

66 European Commission Communication: Regional Protection Programmes, COM(2005) 388 final, 1.9.2005.

67 European Commission Communication: The establishment of a joint EU resettlement programme, COM(2009)456 final, 2.9.2009.

68 The annual report of the European Commission has stated that the negotiations on the Joint EU Resettlement Programme must reach an "operational and positive end". European Commission, Report from the Commission to the European Parliament and the Council, First Annual Report on Immigration and Asylum (2009) COM(2010)214, 6.5.2010.

69 Randall et al., 2014.

70 Art. 8 Protocol 2 on the application of the principles of subsidiarity and proportionality.

71 The Aliens Act 2005 (Sweden), 716/2005, Chapter 4, Section 2 (3) & Chapter 5, Section 1.

72 The Aliens Act 2004 (Finland) 301/2004, Section 88 (1): "Aliens residing in the country are issued with a residence permit on the basis of a need for protection if the requirements for granting asylum under section 87 are not met but the aliens are in their home country or country of permanent residence under the threat of the death penalty, torture or other inhuman treatment or treatment violating human dignity, or if they cannot return there because of an armed conflict or environmental disaster".

73 See the decision CCPR/C/127/D/2728/2016. Views adopted by the Committee under article 5(4) of the Optional Protocol, concerning Communication No. 2728/2016, 7.1.2020, available from: https://tbinternet.ohchr.org/_layouts/15/treatybodyexternal/Download.aspx?symbolno=CCPR%2fC%2f127%2fD%2f2728%2f2016&Lang=en [accessed 1 March 2020].

References

Ammer, M., Mayrhofer, M., Randall, A. and Salsbury, J., 2014. *TIME TO ACT – How the EU Can Lead on Climate Change and Migration*. Brussels: Heinrich-Böll-Stiftung. Available from: https://eu.boell.org/en/2014/06/12/time-act-how-eu-can-lead-climate-change-and-migration.

Black, R., Arnell, N. W., Thomas, D. and Geddes, A., 2013. Migration, immobility and displacement outcomes following extreme events. *Environmental Science and Policy*, 27, 32–43.

Black, R., Bennett, S. R. G., Thomas, S. M. and Beddington, J. R., 2011. Climate change: Migration as adaptation. *Nature*, 478, 447–449.

Boano, C., Zetter, R. and Morris, T., 2008. *Environmentally Displaced People: Understanding Linkages between Environmental Change, Livelihoods and Forced Migration*, Forced Migration Policy Brief No.1. Oxford: Refugee Studies Centre.

Boswell, C., 2003. The 'External Dimension' of EU immigration and asylum policy. *International Affairs*, 79, 619–638.

Christian Aid, 2007. *Human Tide: The Real Migration Crisis*. A Christian Aid Report. London: Christian Aid, 1–51.

Crisp, J., 2003. *Refugee Protection in Regions of Origin: Potential and Challenges*. Brussels: Migration Policy Institute.

De Haas, H., 2018. *European Migrations: Dynamics, Drivers, and the Role of Policies*. Luxembourg: Publications Office of the European Union. Available from: https://publications.europa.eu/en/publication-detail/-/ publication/0e56c014–3232–11e8-b5fe-01aa75 ed71a1.

Foresight, 2011. *Migration and Global Environmental Change*, Final Project Report: Executive Summary. London: The Government Office for Science.

Geddes, A. and Sommerville, W., 2013. *Migration and Environmental Change: Assessing the Developing European Approach*, Policy Briefing Series Issue No. 2. Brussels: Migration Policy Institute.

Gleick, P.H., 2014. Water, Drought, Climate Change, and Conflict in Syria. *American Meteorological Society*, 6, 331–340.

Guy Goodwin-Gill, 1978. *International Law and the Movement of Persons between States*. Oxford: Clarendon Press.

Höing, N. and Razzaque, J., 2012. Unacknowledged and Unwanted? 'Environmental Refugees' in Search of Legal Status. *Journal of Global Ethics*, 8(1), 19–40.

IDMC, 2019. *Disaster Displacement. A Global Review, 2008–2018*. Geneva: IDMC. Available from: https://www.internal-displacement.org/sites/default/files/publications/documents/201905-disaster-displacement-global-review-2008-2018.pdf.

Intergovernmental Panel on Climate Change (IPCC), 1998. *The Regional Impacts of Climate Change. An Assessment of Vulnerability*. Cambridge: Cambridge University Press.

IOM, 2007. *Discussion Note: Migration and the Environment*, Ninety-fourth session. MC/INF/288, Geneva: International Organization for Migration.

IOM, 2011. *Glossary on Migration*. Geneva: International Organization for Migration.

IOM, UNCCD. and Vigil, S., 2019. *Addressing the Land Degradation – Migration Nexus: The Role of the United Nations Convention to Combat Desertification*. Geneva: International Organization for Migration.

Kelley, C.P., Mohtadi, S., Cane, M.A., Seager, R. and Kushnir, Y., 2015. Climate Change and the Recent Syrian Drought. *Proceedings of the National Academy of Sciences*, 112(11), 3241–3246.

Kraler, A., Cernei, T. and Noack, M., 2011. *"Climate Refugees" Legal and Policy Responses to Environmentally Induced Migration*. Study prepared by the Policy Department C – Citizens' Rights and Constitutional Affairs European Parliament for the European Parliament's Committee on Civil Liberties, Justice and Home Affairs. Available from: www.europarl.europa.eu/ committees/de/studiesdownload.html?languageDocument=EN&file=60931.

Magniny, V., 1999. *Les réfugiés et l'environnement. Hypothèse juridique à propos d'une menace écologique*. Doctoral dissertation. Paris: University Paris Panthéon-Sorbonne, Department of Law.

Mayrhofer, M. and Ammer, M., 2014. People Moving in the Context of Environmental Change: The Cautious Approach of the European Union. *European Journal of Migration and Law*, 16(3), 389–429.

Norwegian Refugee Council, 2014. *Annual Report*. Oslo: Norwegian Refugee Coucil. Available from: https://www.nrc.no/globalassets/pdf/annual-reports/2014/nrc-annual-report-2014.pdf.

Pascouau, Y., 2013. *EU Immigration Policy: Act Now Before It Is Too Late*. Commentary European Policy Centre, EPC. Available from: https://www.epc.eu/en/Publications/EU-immigration-policy-act-now~1c771c.

Peers, S., 2011. *EU Justice and Home Affairs Law*. Oxford: University Press.

Petrillo, E.R., 2014. Environmental Migration from Conflict Affected Countries: Focus on EU Policy Response. *European Journal of Migration and Law*, 16, 389–429.

Randall, A., Salsbury, J. and White, Z., 2014. *Moving Stories. The Voices of People Who Move in the Context of Environmental Change*. Oxford: Climate Outreach and Information Network.

Rigaud, K., de Sherbinin, A., Jones, B., Bergmann, J., Clement, V., Ober, K., Schewe, J. et al., 2018. *Groundswell: Preparing for Internal Climate Migration*. Washington: World Bank.

Sgro, A., 2008. Towards Recognition of Environmental Refugees by the European Union. *REVUE Asylon(s)*, 6. Ecological Exoduses. Available from: www.reseau-terra.eu/article844.html [Accessed 18 November 2019].

Stern, N. (Editor), 2006. *The Economics of Climate Change: The Stern Review*. Cambridge: Cambridge University Press.

Tripathi, S., 2017. Climate Refugees: Acknowledging the Existence of an Imminent Threat. *NLUJ Law Review*, 4(1), 21–34.

UNHCR, 2017. *Climate Change, Disasters and Displacement. Guy S Goodwin-Gill and Jane McAdam*. Geneva: Federal Department of Foreign Affairs FDFA. Available from: https://www.unhcr.org/596f25467.pdf.

Vlassopoulos, C., 2012. Climate Change and Migration: Towards a New Nexus for Policy Making in the European Union? In *Human Rights and Climate Change. EU Policy Options*. Brussels: European Parliament, Directorate-General for External Policies of the Union, 50–79.

8 Environmental crime

Assessing and enhancing EU compliance with international environmental law

Maria Marques-Banque

1. Introductory remarks

Although there is no internationally agreed definition of "environmental crime", there is some consensus regarding three of its main characteristics. The first is the diversity of the areas involved (e.g., wildlife crime, illegal logging, pollution crime, illegal mining, waste crime, and illegal, unreported, and unregulated fishing); the second is its progressively transnational character (more than one country may be involved in its execution and effects, or be home to its victims or perpetrators); and the third is its frequent connection with organized crime (involving, in practice, a wide range of associated crimes).

As highlighted by the United Nations (UN), the scale of the problem and the severity of its social, political, and economic impacts "point to the need for the international community to recognize environmental crimes as serious threats to peace and sustainable development, to strengthen the environmental rule of law and adopt and implement coordinated measures to effectively combat environmental crimes".[1]

The European Union (EU) is no stranger to this issue. Data show that the EU is both one of the major waste producers in the world and one of the largest global markets for wildlife trade. These features make it an ideal setting for environmental crime.[2]

In this context, this chapter analyses the degree of the EU's compliance with international law regarding environmental crime. The analysis focuses on the use of criminal law, and particularly on the criminal sanctions that may be imposed on individuals in case of non-compliance with the obligations set out in multilateral environmental agreements (MEAs). With this aim in mind, the chapter first identifies the "penal provisions" in the selected MEAs. Second, section 3 addresses the impact of Directive 2008/99/EC on the protection of the environment through criminal law.[3] Section 4 then refers briefly to the sanctions regime provided for in two other regulations that are relevant for environmental crime (the EU Timber Regulation and the EU IUU Fishing Regulation). Finally, taking into account the evolution of European criminal law, section 5 addresses areas in need of improvement if the EU, to quote the European Commission, is to "lead by example and be an effective international partner in environmental governance".[4] Due to space constraints, the enforcement of legislation on environmental crime in the EU will not be examined here.[5]

2. Multilateral environmental agreements, EU law, and criminal sanctions

The first issue to be examined is the wording of MEAs with regard to sanctions. Certain authors consider that MEAs "set out clear penal prohibitions".[6] However, the wording of the options used in these MEAs suggests three different scenarios:

1   Express reference to criminal sanctions. In very few cases do MEAs expressly refer to the criminal nature of sanctions. One MEA that does is the Basel Convention on the Control of Transboundary Movements of Hazardous Wastes and their Disposal (Basel Convention), whose article 4 unequivocally establishes the criminal nature of the illicit traffic of hazardous and other waste and imposes the obligation on the parties "to take appropriate legal, administrative and other measures to implement and enforce the provisions of this Convention, including measures to prevent and punish conduct in contravention of the Convention".[7]

2   Indirect reference to criminal sanctions. Most MEAs use formulas that only *suggest* the criminal nature of sanctions. This may be done via the reference to their severity and dissuasive capacity (as in the case of the International Convention for the Prevention of Pollution from Ships [MARPOL Convention], which adds that the sanctions "shall be equally severe irrespective of where the violations occur");[8] or via the use of the verb "to penalize" (as in the case of the Convention on International Trade in Endangered Species of Wild Fauna and Flora[9] [CITES Convention] or the Cartagena Protocol on Biosafety to the Convention on Biological Diversity[10] [Cartagena Protocol] – although with less intensity in the latter case, since measures to penalize illegal transboundary movements of living modified organisms are to be adopted by the parties only "if appropriate"). As some authors stress, the fact that the treaties do not define the word "penal" gives the parties the possibility to decide on the nature of the sanctions.[11]

3   Indirect reference to sanctions. Finally, some MEAs only refer to the obligation of the parties to adopt "adequate measures" to guarantee compliance with the obligations derived from the agreements. Of course, such measures may be sanctioning measures and may be criminal in nature, but it is important to note that, in principle, parties have greater flexibility in deciding their strategy. This is the case, for example, of the Rotterdam Convention on the Prior Informed Consent Procedure for Certain Hazardous Chemicals and Pesticides in International Trade (Rotterdam Convention), which, unlike CITES, does not expressly suggest sanctions in specifying the measures to be adopted;[12] of the Stockholm Convention on Persistent Organic Pollutants (Stockholm Convention), which refers generically to the "legal and administrative measures" necessary to eliminate the production or limit the import

or export of certain chemicals;[13] of the Montreal Protocol on Substances that Deplete the Ozone Layer (Montreal Protocol) to which the provisions of the Vienna Convention apply, i.e., the application of "appropriate legislative or administrative measures" to control, limit, reduce, or prevent certain activities, without expressly referring to "sanctions";[14] of the Convention on Biological Diversity, which refers to the obligation of the parties to "develop or maintain necessary legislation and/or other regulatory provisions for the protection of threatened species and populations";[15] and the Bern Convention on the Conservation of European Wildlife and Natural Habitats of the Council of Europe (Bern Convention), which refers to the obligation of taking "appropriate and necessary legislative and administrative measures" in several different articles.[16]

The second issue to be considered is the series of limitations deriving from the evolution of European criminal law and from the relationship between EU law and international law. The regulations regarding the prevention of pollution of the marine environment from ships serve as an example in both cases.

As is known, this issue was regulated in 1973 by the MARPOL Convention, which came into force in 1983 after the signing of the Protocol of the same name in 1978. This international agreement imposes the obligation of the parties to provide for penalties "adequate in severity to discourage violations of the present Convention", which have been interpreted as criminal sanctions. At the EU level, the issue was regulated by Directive 2005/35/EC of the European Parliament and of the Council on ship-source pollution and on the introduction of penalties for infringements.[17] In its recital number 3, the Directive referred to the need to harmonize the application of the MARPOL Convention in the member states, since significant differences could be observed in the practices of member states relating to the imposition of penalties for discharges of polluting substances from ships. As supplementary regulations, in recital 6 the Directive referred to the detailed rules on crimes and sanctions established in Council Framework Decision 2005/667/JHA of 12 July 2005 to strengthen the criminal law framework for the enforcement of the law against ship-source pollution.[18] However, this Council Framework Decision was annulled by the European Union Court of Justice (EUCJ) in October 2007.[19] In order to fill the legal vacuum following the judgment, the European Parliament and of the Council passed Directive 2009/123/EC of 21 October 2009.[20] In its first recital, Directive 2009/123/EC repeats that its objective is "to approximate the definition of ship-source pollution offences committed by natural or legal persons, the scope of their liability and the criminal nature of penalties that can be imposed for such criminal offences by natural persons".

Let us see how this case illustrates some of the difficulties that have arisen in the EU's compliance with international environmental law. Firstly, we must refer to the EU's competences in criminal matters prior to the Lisbon Treaty. The Council Framework Decision 2005/667/JHA was annulled after the EUCJ judgement of 13 September 2005, which resolved a serious institutional conflict between the

European Commission and the Council of the European Union in the field of environmental crime.[21] The EUCJ clarified the distribution of competences in criminal matters between the then first pillar (also called "community pillar" and made up of the constituent treaties), and the third pillar (related to police and judicial cooperation in criminal matters). In summary, the judgment affirmed the possibility of obliging member states to foresee criminal sanctions within the framework of the first pillar, when this is necessary to guarantee the effectiveness of community law. This relevant judgment resulted in the annulment of Council Framework Decision 2003/80/JHA of 27 January 2003, on the protection of the environment through criminal law[22] and impacted on the Council Framework Decision 2005/667/JAI on pollution from ships, causing its subsequent annulment.[23]

From the point of view of the approximation of criminal sanctions, it should be noted that Council Framework Decision 2005/667/JHA obliged member states to provide for effective, proportionate, and dissuasive criminal sanctions in their national legislation. It also specified the typology and level of sanctions, establishing minimum and maximum limits in relation to custodial sentences, and even in relation to the fines to be imposed on legal persons (although in this case, it left the member states the power to decide the criminal or administrative nature thereof). On the basis of the European legal framework then in force, the EUCJ considered that the setting of the typology and level of sanctions was in excess of the EU's criminal powers. As a consequence, Directive 2009/123/EC, currently in force, only imposes the obligation to adopt the necessary measures to guarantee that the offences referred to in articles 4 and 5 of the Directive are considered criminal offences and are punishable by effective, proportionate, and dissuasive sanctions.

Considering that the Directive sets out the obligation to provide for criminal sanctions, it might be concluded that it formally complies with the MARPOL Convention. However, as will be examined later in relation to other European regulations, the degree of flexibility available to member states when defining sanctions leads to significant disparities between them and compromises the effective protection of the environment. Thus, it will be particularly interesting to assess the room for improvement existing since the entry into force of the Lisbon Treaty.

Secondly, the regulations on pollution from ships also illustrate the difficulties arising from the relationship between EU law and international law. As Pereira recalls, the EUCJ had the opportunity to rule on the compatibility of Directive 2005/35/EC and international law in the Intertanko case.[24],[25] Despite the express reference of the Directive to the MARPOL Convention, the court refused to analyse the legality of the Directive in the light of the Convention. The court declared that this was not possible since the community, by itself, was not a signatory party to the Convention, even though the member states were. As Pereira continues to point out, it is especially relevant that the court did not examine the compatibility of the Directive with the Convention on the Law of the Sea (UNCLOS). In this case, although the European community was a signatory of the Convention, the court considered that UNCLOS had no direct effect on the European legal order, as it did not aim to confer rights on individuals or legal persons such as shipping companies. The conclusion reached by this author, relevant to the subject at hand here, is that the Intertanko judgment suggests a desire for caution on the part of the court when assessing the legality of secondary European legislation on environmental crime, in the light of the MEAs, "in order to maintain the delicate balance between international law and the law of the European Union and preserve the position of the EU when negotiating and adopting such international agreements".[26]

3. The Directive 2008/99/EC on the protection of the environment through criminal law

Taking the aforementioned considerations into account, this section examines to what extent the EU currently complies with the sanctions regime set out in the MEAs, and the role of Directive 2008/99/EC on the protection of the environment through criminal law.

Directive 2008/99/EC, passed after the 2005 annulment of EU Council Framework Decision 2003/80/JHA, aims to address the rise in environmental offences with transboundary effects, in a context of non-complete compliance with the laws for the protection of the environment. Article 3 of Directive 2008/99/EC describes nine serious conducts that member states should consider criminal in their national legislation, when they are unlawful and are committed intentionally or, at least, by serious negligence. For the purposes of this Directive, "unlawful" is understood to be an infringement of community legislation specified in the Annexes to the Directive, or of a law, an administrative regulation of a member state, or a decision taken by a competent authority of a member state that gives effect to the aforementioned community legislation. Article 4 sets out that member states shall ensure that inciting, aiding, and abetting the intentional conduct referred to in article 3 is punishable as a criminal offence. As regards sanctions, article 5 sets out that member states shall take the necessary measures to ensure that the offences referred to in articles 3 and 4 are punishable by effective, proportionate, and dissuasive criminal penalties. When it comes to the liability of legal persons, sanctions must be effective, proportionate and dissuasive, leaving member states to decide whether the offences are criminal or administrative in nature.[27]

The analysis will focus on the MEAs referred to previously (except for the already mentioned MARPOL Convention), classified according to their degree of "criminalization". In Table 8.1, each of the MEAs can be compared with the sanctions regime set out in the most relevant secondary European legislation. The third column shows whether there is a specific offence in Directive 2008/99/EC and whether the related European legislation was included in the Annex (as mentioned previously, this is relevant to the interpretation of the term "unlawful").

Table 8.1 shows that European environmental legislation does not set out express criminal provisions even in the clearest case of the "international obligation of criminalization" (Basel Convention). In most cases, and due to the limitations noted previously, the current European legislation opts for the formula "effective, proportionate and dissuasive sanctions", widely used since the Greek

Table 8.1 MEAs, sanctions, and Directive 2008/99/EC

MEAs with express reference to criminal sanctions

	Sanctions in EU legislation	Directive 2008/99/EC
Basel Convention	effective, proportionate, and dissuasive[28]	Offence art. 3c Legislation in Annex

MEAs with indirect reference to criminal sanctions

	Sanctions in EU legislation	Directive 2008/99/EC
CITES	appropriate to the nature and gravity of the infringement[29]	Offence art. 3g Legislation in Annex
Cartagena Protocol (Convention on Biological Diversity)	effective, proportionate, and dissuasive[30]	Annex: Only Directive 2001/18/EC

MEAs with indirect reference to sanctions

	Sanctions in EU legislation	Directive 2008/99/EC
Rotterdam Convention	effective, proportionate, and dissuasive[31]	Regulation in force passed after the Directive. Annex: absence of the previous legislation regarding the export and import of dangerous chemical products
Stockholm Convention	effective, proportionate, and dissuasive[32]	Offences art. 3b and d Regulation in force passed after the Directive. Annex: previous legislation with same provision regarding sanctions
Montreal Protocol (Vienna Convention)	effective, proportionate and dissuasive[33]	Offence art. 3i Regulation in force passed after the Directive. Annex: previous legislation with same provision regarding sanctions
Convention on Biological Diversity	No mention of sanctions, only of appropriate or necessary measures[34]	Offences art. 3f and 3h Annex: Birds Directive and Habitats Directive
Bern Convention (Council of Europe)	No mention of sanctions, only of appropriate or necessary measures[35]	Offences art. 3f and 3h Annex: Birds Directive and Habitats Directive

maize case, which leaves member states the decision about the nature of the sanctions to apply.[36]

Directive 2008/99/EC therefore had a direct impact on the criminalization of non-compliance with MEAs by obliging member states to use criminal law to strengthen compliance with the related European environmental legislation. The impact is especially relevant in cases in which the sanctions in the international and European mandates were particularly weak (e.g., in the Convention on Biological Diversity and the Bern Convention). The Directive defines specific crimes related to the protection of flora, fauna, and habitats, and even extends its protection to include serious negligence.

However, Directive 2008/99/EC also has its shortcomings. The most significant gap concerns the Rotterdam Convention. The Directive does not include an offence that specifically refers to the export and import of dangerous chemical products; nor does it include, in the Annex, the relevant European legislation on the matter that was in force at the time of its approval. With regard to the Cartagena Protocol, there is also a striking gap: the Annex to the Directive omits Regulation (EC) n°. 1946/2003 regarding the transboundary movement of genetically modified organisms.

However, when assessing the impact of Directive 2008/99/EC, it is important to underline that it has been subject to significant criticism both for substantive reasons (the use of criminal law in this area) and for formal reasons (for example, the legislative technique used to describe the offences).[37] To these criticisms, it is worth adding the results of the analysis of its implementation in the national legislations of EU member states. In this respect, with regard to criminal penalties, the main conclusion that emerges from a comparison of its implementation by member states is that, although a high degree of compliance is observed in terms of the use of criminal law, there are significant disparities in relation to the type and severity of the penalties.[38] This finding, which might be seen as a deficit of proportionality between member states, is, however, the consequence of another problem regarding the approximation of criminal sanctions in the EU: the principle of proportionality must operate, in the first instance, in the context of each national criminal justice system.

4. The protection of the environment through criminal law in the EU beyond Directives 2008/99/EC and 2009/123/EC

Directives 2008/99/EC and 2009/123/EC set out, for the first time, the obligation to use criminal law to protect the environment. However, the protection of the environment through criminal law in the EU is not limited to these directives; there are other EU regulations that indirectly open the door to the use of criminal sanctions in areas of special environmental relevance, in which the EU, as an important destination market, should play a leading role in terms of international responsibility. Here, I will refer very briefly to the sanctions regime for two of these regulations: the European Timber Regulation (EUTR)[39] and the European Regulation on Illegal, Unreported and Unregulated Fishing (IUU).[40]

Regulation (EU) No. 995/2010 of the European Parliament and of the Council of 20 October 2010, laying down the obligations of operators who place timber and timber products on the market (EUTR), was adopted in the framework of the European Union Program for Forest Law Enforcement, Governance and Trade (FLEGT).[41] The scale of the problem of deforestation and its relationship with climate change and the loss of biodiversity led the EU to develop specific action to tackle illegal logging and the trade associated with this practice.

To this end, the EUTR set out a series of obligations for operators who commercialize timber and timber products, including the prohibition of placing illegally harvested timber or timber products deriving from such timber on the EU market. Regarding sanctions, it set out the obligation of the member states to provide for "effective, proportionate and dissuasive" penalties in case of non-compliance with the prohibitions, including the corresponding fine, the seizure of the timber, and the suspension of the authorization to trade.

The comparative analysis of the implementation of the EU Timber Regulation in the 28 member states (the UK being included at the time of the analysis) and Norway reveals, once again, significant differences. Member states opted mainly for a combination of criminal and administrative sanctions (13 member states and Norway), or for administrative sanctions (11 member states).[42] Only two solely established criminal sanctions. When provided, the maximum custodial sentences range from one month to six years, being between one and three years in most member states.[43] Regarding fines (criminal or administrative), the minimum fine is 14 euros and there is no limit on the maximum. The highest fines are those imposed for placing illegally harvested timber and timber products on the EU market.[44] Significantly, in 2018, the European Parliament asked the European Commission to extend the scope of Directive 2008/99/EC to include illegal timber logging.[45]

As regards the European Regulation on Illegal, Unreported and Unregulated Fishing (IUU), its article 44 obliges member states "to ensure that a natural person having committed or a legal person held liable for a serious infringement is punishable by effective, proportionate and dissuasive administrative sanctions". In order to guarantee the dissuasive nature of the sanctions and to seek greater harmonization, it regulates the corresponding fine, suggests minimum amounts, and lays down that "in applying these sanctions, the member states shall also take into account the value of the prejudice to the fishing resources and the marine environment concerned". The same article 44 establishes in the last paragraph that: "Member States may also, or alternatively, use effective, proportionate and dissuasive criminal sanctions".

In this case, it has not been possible to conduct a comparative analysis of the implementation of the IUU Regulation in member states. The IUU Regulation obliges member states to report to the European Commission on the application of the Regulation, and the European Commission to present a report every three years to the European Parliament and the Council. However, this information has not been made public, thus making it difficult to analyse the dissuasive, effective, and proportionate nature of the sanctions in this field.

However, the terms of the public consultation on Directive 2008/99/EC carried out by the European Commission in late 2019 are indicative: "Currently, breaches of EU fisheries legislation are generally not criminalized. Do you find it justified and coherent that breaches of fisheries legislation should be criminalized?" Without offering a comprehensive analysis of all the member states, the few existing comparative studies on sanctions on IUU fishing in the EU confirm the low level of criminalization in this area, as well as significant differences between the member states regarding the amounts imposed as fines.[46]

Moreover, despite the efforts of the United Nations Food and Agriculture Organization (FAO) to tackle IUU fishing,[47] the lack of a comprehensive approach to fisheries crimes operates as an aggravating factor. As Stølsvik highlights, despite the progressive international recognition of IUU fishing as transnational organized crime, "no single international organization has a mandate to cover both fisheries management issues and crime in the fisheries sector".[48]

5. A proposal to enable the EU to reinforce its position within the international environmental regime

In European criminal law, some authors have stressed the need to strengthen the EU's external dimension in environmental crime. For Vervaele, in addition to protecting common interests, European criminal policy must be based on common values such as the prohibition of the death penalty or torture. This author also suggests that the EU should consider the protection of Mother Earth and the environment as a legitimating common value for European criminal justice.[49]

As regards environmental crime, the analysis carried out in this chapter points to two ways in which the political influence of the EU could be reinforced and its regulatory power over international environmental governance strengthened. The first refers to actions affecting EU criminal law which have an impact on compliance with international environmental law. The second refers to the external dimension of the EU, not only as a recipient of international legislation, but as a key political actor in its drafting and negotiation.

The implementation of the European legislation analysed so far suggests the presence of major gaps and a low level of harmonization of sanctions in relation to environmental crimes, which has a negative impact on the effective protection of the environment and compliance with international law. Therefore, it is particularly important to establish whether there is room for improvement.

The main action to consider is the one most debated by scholars: the revision of Directive 2008/99/EC, which, at the time of writing, is under evaluation by the European Commission.

Since the Lisbon Treaty came into force in 2009, there have been two ways of establishing minimum criminal rules using Directives.[50] The first, foreseen in article 83(1) of the Treaty on the Functioning of the European Union (TFEU),[51] was already present in the Treaty of Amsterdam (1997) and corresponds to what is known as "security criminalization". This concerns the possibility that the European Parliament and the Council may establish minimum rules regarding the definition of criminal offences and sanctions in the areas of particularly serious crime with a cross-border dimension resulting from the nature or impact of such offences, or from a special need to combat them on a common basis. The TFEU provides an appraised list of crimes (terrorism, trafficking in human beings, sexual exploitation of women and children, illicit drug trafficking, illicit arms trafficking, money laundering, corruption, counterfeiting of means of payment, computer crime, and organized crime), albeit with an open clause that allows the addition of other criminal areas.

Article 83.2 of the TFEU refers to "functional criminalization". The result of a long evolution, this article reflects the jurisprudence deriving from the institutional conflict in the field of environmental crimes. It provides for the possibility of using Directives to establish minimum rules with regard to the definition of criminal offences and sanctions, when the approximation of the laws and regulations of member states in criminal matters is essential to ensure the effective implementation of an EU policy in an area that has been subject to harmonization measures.

These options have already been used in other areas. Based on articles 83(1) or 83(2) of the TFEU, Directives have already been approved which, for certain cases, establish maximum penalties of deprivation of liberty for at least five or ten years (trafficking in human beings) and four years (money laundering and fraud affecting the financial interests of the EU).[52] It should be noted that, with regard to harmonization, scholars have already underlined the limited effect of establishing a minimum for maximum penalties, but not for minimum penalties. However, this last option, theoretically possible under article 83 of the TFEU, is still far from being fully accepted by member states, despite the arguments put forward in its favour (namely, its greater deterrent effect, the prevention of forum shopping, and greater use of international cooperation mechanisms).[53]

Therefore, the possibility exists of revising the Directive and intensifying the protection of the environment through criminal law. The revision of the Directive would have benefits, such as the possibility of: 1) including or reinforcing areas of transnational environmental crime that are not yet present or are insufficiently addressed (for example, illegal timber logging and trading, IUU fishing, and crime related to the Rotterdam Convention and to the Cartagena Protocol); 2) improving the legislative technique used in the description of offences; and 3) reformulating the sanctions framework, especially in relation to organized environmental crime. In this context, it should be remembered that although a full harmonization of sanctions will never be possible on the basis of article 83 of the TFEU, establishing minimum levels for maximum penalties has a direct effect on the applicability of the United Nations Convention against Transnational Organized Crime[54] (UNTOC). UNTOC applies only to "serious crimes" as defined in article 2 of the Convention – that is, offences punishable by a maximum deprivation of liberty of at least four years, or a more serious penalty.

The revision of the Directive is also an opportunity to consider the Council of Europe Convention on the Protection of the Environment through Criminal Law (not in force).[55] Provisions establishing more stringent protection could be considered in order to improve the Directive.[56] However, it is important to differentiate between cases in which the use of criminal law is justified (such as transnational organized environmental crime), and cases in which sanctions of a different nature could be equally effective, dissuasive, and proportionate. But for this – we should also stress – the EU must establish mechanisms to make the relevant information on the use of environmental sanctions readily available.

It is still important to underline that a greater use of criminal law can only be regarded as an improvement if we accept that criminal law is the best strategy for the effective protection of the environment. In the criminal law literature, however, for many reasons that cannot be discussed in depth here, there is no consensus regarding the priority use of criminal law or regarding the criminal nature of sanctions as if they were the only actions with dissuasive capacity in this area. Thus, while some authors advocate extending Directive 2008/99/EC based on the legal framework deriving from the Lisbon Treaty,[57] others warn of the risk of over-criminalization in the environmental field because of the existence of "too much harmonization that is too detailed".[58] Consistent with the European Parliament Resolution of 22 May 2012 on an EU approach to criminal law,[59] scholars insist on the need to use criminal law only as a last resort and to explore alternatives to criminal sanctions (in what is known as the "toolbox approach").[60] Likewise, in order to make better-founded proposals for the sanctions regime, scholars stress the need for more criminological studies, studies carried out from the perspective of the economic analysis of the law, and, again, the need for a greater systematization and access to empirical data regarding the enforcement of environmental crimes.[61]

Still, considering the first way in which the EU could strengthen its political influence, and as a consequence of the compliance deficit observed in this area, attention has been drawn to the importance of the instruments adopted within the framework of the areas of freedom, security, and justice to facilitate cooperation in criminal matters. Some of these instruments (such as the European arrest warrant, the European investigation order, and the Eurojust and Europol agencies) are essential for the effective implementation of Directives 2008/99/EC and 2009/123/EC.[62] In this regard, a coordinated EU strategy on environmental crime must bear in mind that the applicability of most of these instruments depends on the establishment of minimum requirements of seriousness of the offences.[63] As pointed out in the literature, in the future, the EU should also consider extending the powers of the European Public Prosecutor's Office to include serious environmental crime with cross-border effects.[64]

Turning now to the second way in which it might extend its influence, the EU should have a leadership role, at least, in an area in which there is a greater consensus on the need for coordinated and global action: transnational organized environmental crime.[65] As highlighted in the framework of the EFFACE research project, "environmental crime should be considered a serious crime when it comes to organized crime and should also be considered a crime determining other serious crimes such as money laundering and corruption".[66]

The measures that have been suggested here are the ones that have an impact on the criminal sanctions framework of the EU member states, which is the focus of this chapter. Thus, with regard to the external dimension of EU action to fight against environmental crime, it should be noted that international law scholars have analysed the issue and formulated policy recommendations addressing governance and enforcement.[67]

Along with intensifying international coordination and cooperation mechanisms, more direct EU action has also been called for with regard to the possibility of adopting an additional protocol to the United Nations Convention against trans-national organized crime.[68] It is also essential to intensify the role of customs in the fight against transnational environmental crime and to increase international cooperation in this area, particularly in relation to the MEAs covered by the Green Customs Initiative (among others, CITES, the Basel Convention, the Montreal Protocol, the Stockholm Convention, the Rotterdam Convention, and the Cartagena Protocol).[69]

6. Final remarks

This chapter has addressed the EU's compliance with international environmental law regarding the protection of the environment through criminal law and the action that the EU might take in order to improve its implementation at a regulatory level.

The analysis of a set of relevant MEAs and of the related European legislation shows the impact of Directives 2008/99/EC and 2009/123/EC on the criminalization of non-compliance with MEAs. This impact is especially relevant in cases in which the international and European mandates are particularly weak in terms of the imposition of sanctions.

However, with regard to environmental crime, the EU can do more. Along with the possible revision of Directive 2008/99/EC, whose complexity and limitations derive from the evolution of European criminal law and from the scholarly debate on the subject, there is an urgent need to take regulatory measures to overcome the deficit of enforcement in this area. This will require a comprehensive vision of the impact of the sanctions regime on the instruments adopted within the framework of the areas of freedom, security, and justice in order to facilitate cooperation in criminal matters.

The external dimension of the EU should also be emphasized in the fight against transnational organized environmental crime, either by promoting the process of negotiation and approval of international legislation in this regard, or by intensifying legislative mechanisms for international cooperation. Special attention must be paid to the key role of customs in preventing the import and export from the EU of products whose illegal trade has serious global, environmental, political, economic, and social impacts.

Notes

1 UNEP, 2018, p. VIII.
2 UNEP, 2015; TRAFFIC, 2014.
3 OJ No. L 328, 6.12.2008.
4 EU Commission Communication: EU Actions to improve environmental compliance and governance, COM(2018)10, p. 8.
5 Case studies were compiled as part of European Union Action to Fight Environmental Crime (EFFACE) Research Project. EFFACE was an EU-funded research project involving 11 European research institutions and think tanks. As stated on their webpage, EFFACE assessed the impacts of environmental crime as well as effective and

feasible policy options for combating it from an interdisciplinary perspective, with a focus on the EU. Its findings and reports are available from: http://efface.eu/ [accessed 20 March 2020].

6 Regarding the Basel Convention, CITES and MARPOL, Mégret, 2010, p. 19; Pereira, 2015b, pp. 130–131.
7 UNTS, vol. 1673, p. 57. Of another opinion Mitsilegas et al., 2015, p. 85: "The wording of Article 4(3) does not impose a clear obligation to make illegal traffic criminal, as it simply says that parties 'consider' it to be criminal".
8 UNTS, vol. 1340, p. 61 and 1341, p. 3.
9 UNTS, vol. 993, p. 243.
10 UNTS, vol. 2226, p. 208.
11 Cho, 2000/2001, p. 21.
12 UNTS, vol. 2244, p. 337.
13 UNTS, vol. 2256, p. 119.
14 UNTS, vol. 1522, p. 3.
15 UNTS, vol. 1760, p. 79.
16 Council of Europe: ETS No. 104.
17 OJ No. L 255, 30.9.2005.
18 *Ibid.*
19 Case C-440/05, Commission of the European Communities v Council of the European Union, 23.10.2007. ECLI:EU:C:2007:625.
20 Directive 2009/123/EC of the European Parliament and of the Council of 21 October 2009 amending Directive 2005/35/EC on ship-source pollution and on the introduction of penalties for infringements, OJ No. L 280, 27.10.2009.
21 Case C-176/03, Commission of the European Communities v Council of the European Union, 13.09.2005. ECLI:EU:C:2005:542.
22 OJ No. L 29, 5.2.2003.
23 Symeonidou-Kastanidou, 2009; Pigrau and Campins, 2008; Mitsilegas et al., 2016, pp. 279–283; Pereira, 2015b, pp. 24–36, 186–198.
24 Pereira, 2015a, p. 265.
25 Case C-308/06, International Association of Independent Tanker Owners (Intertanko) and Others v Secretary of State for Transport, 03.062008. ECLI:EU:C:2008:312.
26 Pereira, 2015a, p. 267; Mitsilegas et al., 2016, pp. 288–291.
27 See the reasons in Pereira, 2015b, pp. 267–273.
28 Art. 50 Regulation (EC) No 1013/2006 of the European Parliament and of the Council of 14 June 2006 on shipments of waste, OJ No. L 190, 12.7.2006.
29 Art. 16.2 Council Regulation (EC) No 338/97 of 9 December 1996 on the protection of species of wild fauna and flora by regulating trade therein, OJ No. L 61, 3.3.1997.
30 Art. 33 Directive 2001/18/EC of the European Parliament and of the Council of 12 March 2001 on the deliberate release into the environment of genetically modified organisms and repealing Council Directive 90/220/EEC, OJ No. L 106, 17.04.2001, and art. 18 Regulation (EC) No 1946/2003 of the European Parliament and of the Council of 15 July 2003 on transboundary movements of genetically modified organisms, OJ No. L 287, 5.11.2003.
31 Art. 28 Regulation (EC) No 649/2012 of the European Parliament and of the Council of 4 July 2012 concerning the export and import of hazardous chemicals (recast), OJ No. L 201, 27.7.2012.
32 Art. 14 Regulation (EC) No 2019/1021 of the European Parliament and of the Council of 20 June 2019 on persistent organic pollutants (recast), OJ No. L 169, 25.6.2019.
33 Art. 29 Regulation (EC) No 1005/2009 of the European Parliament and of the Council of 16 September 2009 on substances that deplete the ozone layer (recast), OJ No. L 286, 31.10.2009.
34 Art. 6.2 Council Directive 92/43/EEC of 21 May 1992 on the conservation of natural habitats and of wild fauna and flora, OJ No. L 206, 22.7.1992; Directive 2009/147/EC of the European Parliament and of the Council of 30 November 2009 on the conservation of wild birds, OJ No. L 20, 26.1.2010.

35 *Ibid.*
36 Case C-68/88, Commission of the European Communities v Hellenic Republic, 21.09.1989. ECLI:EU:C:1989:339.
37 Faure, 2017b, pp. 346–349; Vagliasindi, 2012; Faure, 2011.
38 Maria Marques-Banque, 2018; Perilongo and Corn, 2017; Pereira, 2017.
39 Council Regulation (EU) No 995/2010 of the European Parliament and of the Council of 20 October 2010 laying down the obligations of operators who place timber and timber products on the market, OJ No. L 295, 12.11.2010.
40 Council Regulation (EC) No 1005/2008 of 29 September 2008 establishing a Community system to prevent, deter and eliminate illegal, unreported and unregulated fishing, amending Regulations (EEC) No 2847/93, (EC) No 1936/2001 and (EC) No 601/2004 and repealing Regulations (EC) No 1093/94 and (EC) No 1447/1999, OJ No. L 286, 29.10.2008.
41 EU Commission Communication: Forest Law Enforcement, Governance and Trade (FLEGT) – Proposal for an EU Action Plan, COM(2003)25.
42 For two Member States (Denmark and the Republic of Croatia), additional information was requested but no response was obtained.
43 Maria Marques-Banque, 2019. The study was finalized in 2019.
44 UNEP-WCMC, 2018.
45 European Parliament: Resolution of 11 September 2018 on transparent and accountable management of natural resources in developing countries: the case of forests (2018/2003(INI)), P8_TA (2018)0333, n84.
46 European Parliament. Directorate-General for Internal Policies. Policy Department B: Structural and Cohesion Policies, 2014, p. 46. As a case study, particularly interesting is the judgment of the Spanish Supreme Court no. 5654 of 23 December 2016 (ECLI:ES:TS:2016:5654) on a case of illegal fishing in international marine waters. See an analysis in Oanta, 2019; Pons Ràfols, 2017.
47 Pereira, 2015b, p. 254.
48 Stølsvik, 2019, p. 127.
49 Vervaele, 2019, p. 10.
50 Mitsilegas, 2014; Miglietti, 2014; Pereira, 2015b, pp. 202–206.
51 OJ No. C 326, 26.10.2012.
52 On the basis of art. 83(1) of the TFUE, the Directive 2011/36/EU of the European Parliament and of the Council of 5 April 2011 on preventing and combating trafficking in human beings and protecting its victims, and replacing Council Framework Decision 2002/629/JHA, OJ No. L 101, 15.4.2011 and the Directive (EU) 2018/1673 of the European Parliament and of the Council of 23 October 2018 on combating money laundering by criminal law, OJ No. L 284, 12.11.2018. On the basis of art. 83(2) of the TFUE, Directive (EU) 2017/1371 of the European Parliament and of the Council of 5 July 2017 on the fight against fraud to the Union's financial interests by means of criminal law, OJ No. L 198, 28.7.2017.
53 De Bondt and Miettinen, 2015; De Bondt, 2014; Pereira, 2015b, pp. 279–280.
54 UNTS, vol. 2225, p. 209.
55 Council of Europe: ETS No. 172.
56 See an analysis of the Convention of the Council of Europe and a comparison with the Directive 2008/99/EC in Pereira, 2015a.
57 Vagliasindi, 2012, p. 330.
58 Vervaele, 2019, p. 11.
59 P7_TA (2012)0208.
60 Faure, 2017a; Pereira, 2015b, pp. 89–90, 281–288, 292–317; Faure and Svatikova, 2012; Faure, 2004.
61 Faure, 2017b, pp. 338–342; Perilongo and Corn, 2017, p. 250.
62 Pereira, 2017, pp. 161–162; Vervaele, 2019, p. 12; Mitsilegas and Giuffrida, 2017.
63 De Bondt, 2014, pp. 155–156.
64 Di Francesco Maesa, 2018; De Angelis, 2019.
65 On 21 November 2016, the United Nations General Assembly recognized, for the first time, environmental crime as a type of transnational organized crime. See United

Nations General Assembly, Resolution 71/19 on the Cooperation between the United Nations and INTERPOL, A/RES/71/19.
66 Fajardo, 2016, p. 30.
67 Fajardo, 2016; Cardesa-Salzmann, 2016.
68 Global Initiative Against Transnational Organized Crime, 2015; of the same opinion Vervaele, 2019, p. 14.
69 "The Green Customs Initiative is a partnership of international organisations and secretariats that cooperate to enhance the capacity of customs and other relevant border enforcement personnel to deal with trade in 'environmentally-sensitive items'" (Green Customs, p. 1). Its legal basis is the Decision 21/27 of the UNEP Governing Council on Compliance with and enforcement of multilateral environmental agreements.

References

Cardesa-Salzmann, A., 2016. Multilateral Environmental Agreements and Illegality. In Elliot, L. and Schaedla, W.H. (Editors), *Handbook of Transnational Environmental Crime*. Cheltenham: Edward Elgar Publishing, 299–321.

Cho, B-S., 2000/2001. Emergence of an International Environmental Criminal Law? *UCLA Journal of Environmental Law and Policy*, 19(11), 11–47.

De Angelis, F., 2019. The European Public Prosecutor Office (EPPO). Past, Present and Future. *In Eucrim: The European Criminal Law Associations' Forum*, 4, 272–276.

De Bondt, W., 2014. The Missing Link between Necessity and Approximation of Criminal Sanctions in the EU. *European Criminal Law Review*, 4(2), 147–168.

De Bondt, W. and Miettinen, S., 2015. Minimum Criminal Penalties in the European Union: In Search of a Credible Justification. *European Law Journal*, 21(6), 722–737.

Di Francesco Maesa, C., 2018. EPPO and Environmental Crime: May the EPPO Ensure a More Effective Protection of the Environment in the EU? *New Journal of European Criminal Law*, 9(2), 191–215.

European Parliament. Directorate-General for Internal Policies. Policy Department B: Structural and Cohesion Policies, 2014. *Illegal, Unreported and Unregulated Fishing: Sanctions in the EU*. Available from: www.europarl.europa.eu/RegData/etudes/STUD/2014/529069/IPOL_STU(2014)529069_EN.pdf [Accessed 20 January 2020].

Fajardo, T., 2016. Contribution to Conclusions and Recommendations on Environmental Crime: The External Dimension. *In Study in the Framework of the EFFACE Research Project*. Available from: https://efface.eu/sites/default/files/publications/EFFACE_conclusions_recommendations_external%20dimension.pdf [Accessed 9 March 2020].

Faure, M.G., 2004. European Environmental Criminal Law: Do We Really Need It? *European Energy and Environmental Law Review*, 13(1), 18–29.

Faure, M.G., 2011. The Implementation of the Environmental Crime Directives in Europe. In Gerardu, J., Grabiel, D., Koparova, M.R., Markowitz, K. y Zaelke, D. (Editors), *9th International Conference on Environmental Compliance and Enforcement*. Washington: INECE, 360–371.

Faure, M.G., 2017a. The Development of Environmental Criminal Law in the EU and Its Member States. *Review of European Comparative & International Environmental Law*, 26(2), 139–146.

Faure, M.G., 2017b. The Revolution in Environmental Criminal Law in Europe. *Virginia Environmental Law Journal*, 35(2), 321–356.

Faure, M.G. and Svatikova, K., 2012. Criminal or Administrative Law to Protect the Environment? Evidence from Western Europe. *Journal of Environmental Law*, 24(2), 253–286.

Global Initiative Against Transnational Organized Crime, 2015. *Tightening the Net: Toward a Global Legal Framework on Transnational Organized Environmental Crime*. Available from: www.unodc.org/documents/congress/background-information/NGO/ GIATOC-Blackfish/GIATOC_-_Tightening_the_Net.pdf [Accessed 20 January 2020].

Green Customs. *The Green Customs Initiative: Capacity Building for Environmental Security*. Available from: www.ippc.int/static/media/files/publication/en/2016/11/The_ Green_Customs_Initiative_for_IPPC.pdf [Accessed 20 January 2020].

Marques-Banque, M., 2018. The Utopia of the Harmonization of Legal Frameworks to Fight Against Transnational Organized Environmental Crime. *Sustainability*, 10(10), 3576. Available from: www.mdpi.com/2071-1050/10/10/3576 [Accessed 20 January 2020].

Marques-Banque, M., 2019. Estrategias sancionadoras en materia de cambio climático: la persecución penal del tráfico ilegal de madera en la Unión Europea y en España. *Revista Catalana de Dret Ambiental*, 10(2), 1–42. Available from: https://revistes.urv.cat/index. php/rcda/article/view/2670 [Accessed 20 January 2020].

Mégret, F., 2010. *The Challenge of an International Environmental Criminal Law*. Available from: https://ssrn.com/abstract=1583610 [Accessed 20 January 2020].

Miglietti, M., 2014. The First Exercise of Article 83(2) TFEU Under Review: An Assessment of the Essential Need of Introducing Criminal Sanctions. *New Journal of European Criminal Law*, 5(1), 5–25.

Mitsilegas, M., Fitzmaurice, M., Fasoli, E. and Fajardo, T., 2015. Analysis of International Legal Instruments Relevant to Fighting Environmental Crime. In *Study in the Framework of the EFFACE Research Project*. London: Queen Mary University of London. Available from: https://efface.eu/sites/default/files/11.EFFACE_Analysis%20of%20Inter national%20Legal%20Instruments_0.pdf [Accessed 9 March 2020].

Mitsilegas, V., 2014. EU Criminal Law Competence after Lisbon: From Securitised to Functional Criminalisation. In Acosta Arcarazo, D. and Murphy, C.C. (Editors), *EU Security and Justice Law. After Lisbon and Stockholm*. Oxford: Hart Publishing, 110–128.

Mitsilegas, V., Fitzmaurice, M. and Fasoli, E., 2016. The Relationship between EU Criminal Law and Environmental Law. In Mitsilegas, V., Bergström, M. and Konstadinides, T. (Editors), *Research Handbook in European Criminal Law*. Cheltenham: Edward Elgar Publishing, 272–293.

Mitsilegas, V. and Giuffrida, F., 2017. The Role of EU Agencies in Fighting Transnational Environmental Crime. *Transnational Crime*, 1(1), 1–150.

Oanta, G., 2019. Spain's Action to Control and Suppress Illegal, Unreported and Unregulated Fishing: Current Status and Future Prospects. *The International Journal of Marine and Coastal Law*, 34(4), 642–667.

Pereira, R.M., 2015a. The External Dimensions of the EU legislative Initiatives to Combat Environmental Crime. *Spanish Yearbook of International Law*, 19, 251–268.

Pereira, R.M., 2015b. *Environmental Criminal Liability and Enforcement in European and International Law*. Leiden: Brill.

Pereira, R.M., 2017. Towards Effective Implementation of the EU Environmental Crime Directive? The Case of Illegal Waste Management and Trafficking Offences. *Review of European Comparative & International Environmental Law*, 26(2), 147–162.

Perilongo, G.F. and Corn, E., 2017. The Ecocrime Directive and Its Translation into Legal Practice. *New Journal of European Criminal Law*, 8(2), 236–255.

Pigrau i Solé, A. and Campins Eritja, M., 2008. La protección penal del medio ambiente a través del Derecho penal en la Unión Europea, a propósito de la Decisión marco 2003/80/JAI del Consejo, de 27 de enero de 2003 y de su anulación: una visión desde el Derecho internacional. In Quintero Olivares, G. and Morales Prats, F. (Editors), *Estudios de Derecho ambiental. Libro Homenaje al profesor Josep Miquel Prats Canut.* Valencia: Tirant lo Blanch, 229–270.

Pons Ràfols, F.X., 2017. Spain and the Fight Against IUU Fishing. *Spanish Yearbook of International Law*, 21, 423–438.

Stølsvik, G., 2019. The Development of the Fisheries Crime Concept and Processes to Address it in the International Arena. *Marine Policy*, 105(July), 123–128.

Symeonidou-Kastanidou, E., 2009. Ship-Source Marine Pollution: The ECJ Judgments and Their Impact on Criminal Law. *European Journal of Crime, Criminal Law and Criminal Justice*, 17(4), 335–358.

TRAFFIC, 2014. *Briefing on Wildlife Trade in the European Union.* Available from: www.traffic.org/general-reports/traffic_pub_gen56.pdf [Accessed 20 January 2020].

UNEP, 2015. *Waste Crime: Waste Risks Gaps in Meeting: The Global Waste Challenge.* Available from: http://web.unep.org/ourplanet/september-2015/unep-publications/waste-crime-waste-risks-gaps-meeting-global-waste-challenge-rapid [Accessed 20 January 2020].

UNEP, 2018. *The State of Knowledge of Crimes that have Serious Impacts on the Environment.* Available from: www.unenvironment.org/resources/publication/state-knowledge-crimes-have-serious-impacts-environment [Accessed 20 January 2020].

UNEP-WCMC, 2018. *Background Analysis of the 2015–2017 National Biennial Reports on the Implementation of the European Union's Timber Regulation (Regulation EU No 995/2010).* Available from: http://ec.europa.eu/environment/forests/pdf/WCMC%20EUTR%20analysis%202017.pdf [Accessed 20 January 2020].

Vagliasindi, G.M., 2012. The European Harmonization in the Sector of Protection of the Environment through Criminal Law: The Results Achieved and Further Needs for Intervention. *New Journal of European Criminal Law*, 3(3–4), 320–331.

Vervaele, J.A.E., 2019. European Criminal Justice in the European and Global Context. *New Journal of European Criminal Law*, 10(1), 7–16.

9 Mainstreaming Sustainable Development Goals into EU policies

Mar Aguilera Vaqués

1. Introductory remarks

Europe is facing a global environmental emergency. Such a reality has been on the table for decades, starting with *silent springs* and the acknowledgement of the *limits of growth* and *beyond*.[1] Out of all the attempts to come up with responses, the idea of sustainable development has been the one to receive the most enduring support. Yet its definition is simple and clear: Sustainable development is development which "meets the needs of current generations without compromising the ability of future generations to meet their own needs".[2] This concept, however, also has a chameleon-like capacity.[3] It has been the leitmotif of thousands of policies and norms all over the world and in the European Union (EU) too. Still, its vague capaciousness has now and again been used by a number of states and several transnational corporations to greenwash unsustainable activities and even to ultimately mislead the public.[4] All the same, the EU has taken a lead in sustainable development policies and made enormous efforts to try to implement it.

This chapter will provide a critical analysis of the main steps taken by the EU to consolidate sustainable development in Europe and globally, and it will examine what the idea of mainstreaming Sustainable Development Goals (SDGs) into EU policies has basically meant for this never-ending path.

After a short introduction, the next section will briefly consider the concept of sustainable development adopted by the European Union. Then section three will focus on the Sustainable Development Goals as policy. Section four will examine the EU as a global trailblazer in sustainable development, and section five will set out an analysis of the progress required in moving from the mainstreaming of the Sustainable Development Goals to the implementation of 2019 European Green Deal. The last section will conclude with some final remarks.

2. From the concept of sustainable development to Sustainable Development Goals

Albeit indirectly, the idea of sustainable development was already present at the 1972 United Nations (UN) Conference on the Human Environment in Stockholm, where the importance of protecting the environment was linked to "the well-being of peoples and economic development throughout the world".[5] However, it was not until the eighties that the concept gained international momentum as a result of renewed environmental concerns relating to the ecological consequences of human activities. Accordingly, the 1987 Report of the UN Commission on Environment and Development, *Our Common Future* (Brundtland Commission), redirected the focus from purely environmental issues to socioeconomic ones. In this regard, the UN came up with the most widely accepted definition of sustainable development, which has been previously noted. The 1987 UN Brundtland Report also issued a call to overcome the fragmented view that addressed development and the environment separately. On one hand, the concept of sustainable development placed an emphasis on needs, particularly those of the most vulnerable; on the other hand, it stated that there were limits to the environment's capacity to meet present and future needs. By and large, the concept introduced an ethical evaluation to the purpose and processes of development.[6]

Nonetheless, concerns arose over the excessively anthropocentric approach of sustainable development, and some critics attacked it as a perpetuation of liberal perspectives on the idea of progress.[7] In addition, carping over the lack of enforcement and efficiency was on the table from the very beginning.[8]

In 1992, the UN Conference on Environment and Development (UNCED) in Rio de Janeiro[9] tried to take a step forward with Agenda 21. UNCED consolidated the trends and established the guiding principles for environmental law and the concepts inherent in the notion of sustainable development "with the goal of establishing a new and equitable global partnership through the creation of new levels of cooperation among States, key sectors of societies and people". As endorsed by the Rio instruments, the concept of sustainable development not only brought about a compromise between development and environmental protection, but also called for an integrative approach to development that considered the importance of equity within the economic system. Indeed, sustainable development became a pillar of Agenda 21. However, Agenda 21 was a non-binding document. The documents issued by the World Summit on Sustainable Development held in Johannesburg in 2002 were not mandatory either.

In 2012, the United Nations Conference on Sustainable Development (Rio+20)[10] insisted once again on the previous agreements and exhorted countries to focus efforts on integrating and ensuring greater consistency among the dimensions of sustainable development and its supporting policies and institutions. Moreover, the Conference again urged countries, albeit with limited success, to eradicate poverty and drastically shift consumption and production patterns.[11] To this end, it called for the setting of goals. These goals were subsumed in the Millennium Development Goals (MDGs – 2000 to 2015), which fell due in 2015, and they streamlined the need for combatting climate change, establishing a new post-2015 development agenda.

By the same token, in 2015, the UN General Assembly Resolution 1/70 set a new agenda for sustainable development to be achieved by 2030. As part of this effort, the UN 2030 Agenda drew up a set of 17 goals,[12] known as the Sustainable Development Goals (SDGs),[13] as well as 169 targets, in a process hailed as widely participative. In September 2015, the SDGs, which were once again non-binding, were adopted by the 193 UN member states of the General Assembly. The SDGs endeavoured to pursue focused and coherent action on sustainable development (Rio+20, para. 246) that could be time-bound and measurable by indicators. It was established that the goals and targets had to include benchmarks that could show evidence of progress on environmental quality, social outcomes, and economic performance. For this to occur, it was determined that states should set objective targets. All in all, the SDGs contained 17 goals, 169 targets, and 304 indicators. However, the SDGs were not free of criticism; some studies, among them a study published by the International Council for Science[14], found that less than a third (29 per cent) of the 169 targets were well defined and consistent with the latest scientific evidence, while another 54 per cent could be more specific, and the remaining 17 per cent were weak.

On the positive side, the 2015 SDGs also offered a roadmap to the world based on economic, environmental, and social pillars. In this vein, the SDGs established a transformational leadership role for some countries, especially in terms of sustainable production and consumption models, given that universality, as the document indicated, "is different from uniformity". Also, while acknowledging that the agenda was equally important to all countries, every nation and region had to define its own institutional arrangements and public policies to accomplish the goals. In this context, the EU played a leading role in the adoption of the 2030 Agenda. It also committed to being a role model for other regions. This accords with the fact that the SDGs reflect EU core values, which are to promote greater social, political, economic, and environmental harmony. The European Green Deal (EGD),[15] which will be discussed later, is the most recent and decisive step taken in this direction.

Lastly, it needs to be mentioned that despite having a definition and a path for sustainable development since the eighties, the concept has not always been clear, specific enough, or free of conflict. For many, all of these sustainable development agreements and goals, in addition to being non-binding, leave too much room for uncertainty.[16] This can be seen not only from a scientific approach – many hold the opinion that the world is not achieving the goals[17] – but also from a conceptual perspective.[18] Proof of the latter point lies in the fact that, in addition to the breadth of interpretations, there were already more than 70 different definitions of sustainable development only five years after Rio.

3. Sustainable Development Goals as an EU policy effort

Following the adoption of the UN 2030 Agenda for Sustainable Development by the EU in 2015, the European Commission convened to establish an overarching European Sustainable Development Strategy with specific objectives, targets, and actions in order to achieve the 17 SDGs. Among others, the 2030 Agenda for Sustainable Development stated that broader measures of progress to complement gross domestic product needed to be developed by the member states and by the EU. Although "harmony with nature" appeared three times in the 2030 Agenda, it is not clear, according to Adelman, how ecological sustainability must be measured. In his opinion, the document is "resolutely anthropocentric and envisages economic growth on a scale incompatible with such harmony" and this means that "[t]he 2030 Agenda contains lofty ambitions and stirring rhetoric but no enforcement mechanisms".[19] No matter how unrealistic, vague, or directionless the 2015 SDGs and the 2030 Agenda directives may be,[20] the truth is that ever since the UN 2030 Agenda for Sustainable Development, the EU has undertaken to integrate the SDGs both in its internal and external policies.[21] Hence, this integration has sought to strengthen environmental considerations and priorities, but it has also diluted them within other norms and policies.[22] The result of these policies is subject to periodic evaluation through objective studies, which indicate the extent of success in implementation. The Statistical Office of the European Union 2019 Report, for instance, proclaims that the results from its "objective assessment . . . show that the EU has achieved progress towards many sustainable development objectives". However, the report also "points to areas where further effort is needed to put the EU on the right track".[23]

As a matter of fact, since the very beginning, EU treaties have been peppered with references to the environment and, ultimately, to sustainable development principles. The preamble to the Treaty on European Union (TEU) already stated that the EU's aim of "economic and social progress" must "take into account the principle of sustainable development and . . . environmental protection". Sustainable development is a fundamental objective of the Union as laid down in article 3.3 of the TEU, and it should play a central role in the debate on, and the narrative for, the future of Europe. Article 3.3 provides that one of the goals of the EU is to aim for a "high level of protection . . . of the environment" but also an "improvement [in its] quality". Indeed, from day one, the environment has been not only an aspect of the EU's economic and social policy, but also a purpose in its own right.[24] In particular, the TFEU contains a specific environmental chapter, Title XX, which pinpoints the guidelines of the Union's environmental policy and confers powers upon the European Parliament and Council to enact legislation to that end. Furthermore, the EU has continuously strived for more inclusive societies built on democracy and the rule of law as reflected in article 2 of the TEU.

It should be borne in mind, however, that among the member states, there was considerable uncertainty about the scope, exact meaning, and legal implications of sustainable development and the SDGs. From the outset, however, it was also widely accepted that the principle of integration was essential for achieving sustainable development and there was a general agreement on the procedural role of law by establishing requirements to include environmental, social, and economic considerations in decision-making processes.

It must also be noted that the initial focus was on the idea that sustainable development should be aimed at regulating the conduct of member states and EU institutions through obligations of means or of best efforts, or due diligence obligations – as opposed to obligations of results.[25] Within the EU context, the primary responsibility for the implementation of Agenda 2030 and its Sustainable Development Goals rested, and still rests, with national governments. This requires the forging of strong partnerships with key actors and stakeholders ranging from local authorities to national institutions, business communities, and, especially, civil society.[26] Given that the 2030 Agenda sets clear, measurable objectives to be achieved by member states and the EU, it implies due diligence obligations such as best efforts, as well as process obligations and an emphasis on the actual result – the outcome. Considering that the obligations are clear, measurable, and well-defined, the approach taken to achieving the goals and targets may be considered as actual obligations of member states – and, thus, binding with the effect of international law – rather than merely political statements.

As an objective, sustainable development must have an influence in the decision-making process of legal subjects. For instance, documents and norms have pointed at some length to the need to substantially integrate sustainable development into the *outcome*, and not only into the *process*, of decisions. This entails nothing less than the mainstreaming of environmental, social, and economic dimensions in all EU policies and better regulation practices. Accordingly, the SDGs have provided, and still provide, specific benchmarks for performance.[27]

A number of tools have been needed to help in measuring the adequacy of the conduct adopted by member states and the EU in the light of their obligation to promote sustainable development. As mentioned earlier, the UN SDGs have endeavoured to pursue focused and coherent action on sustainable development (Rio+20, para. 246) that could be time-bound and measurable by indicators. In that vein, the EU's SDG indicators are an example of similar tools. They were intended to go beyond mere evaluation by contributing to policy formulation and policy shaping, rather than simply serving as a reporting tool. Indeed, the indicators have helped EU policy makers in defining future policies and planning how to better achieve the SDGs. They have also been useful for policy makers to identify deviations in progress towards the SDGs and introduce necessary policy changes in a timely manner to achieve such goals by 2030. However, the use of cross-cutting indicators alone, while valuable, remains insufficient to provide information about synergies and dilemmas among the goals.[28]

All indications are, however, that little will be accomplished without an integrated and holistic SDG implementation strategy. There is little point in establishing SDGs if they cannot be achieved. Despite the announcement of the promising 2019 EGD,[29] which states that "the Commission will work with the Member States to step up the EU's efforts to ensure that current legislation and policies relevant to the Green Deal are enforced and effectively implemented",[30] the annual strategic report on the implementation and delivery of the Sustainable Development Goals regrets that the European Commission has not yet developed an integrated and holistic SDG implementation strategy.[31]

The truth is that, despite all the efforts and important steps, the SDGs have not yet been accomplished.[32] For instance, the 2019 Europe Sustainable Development Report,[33] which was produced by teams of independent experts at the Sustainable Development Solutions Network (SDSN) and the Institute for European Environmental Policy (IEEP)",[34] found that "[w]hile European countries lead globally on the SDGs, none are on track to achieve the Goals by 2030". According to this 2019 ESDR report, only three EU countries are close to achieving the SDGs by 2030, namely Denmark, Sweden, and Finland, while Bulgaria, Romania, and Cyprus rank last among the 28 countries assessed, and the EU as a whole would be 13th. In the ranking, Ireland stands at 14th, while Spain is 15th.

On a similar note, the EU's goal of being a role model for the rest of the world calls for a holistic approach, too.[35] Taking into account that the largest negative impacts are caused by "unsustainable demand for agricultural, forest, and fishery products", the EU should not be an obstacle to others in their attempts to reach the SDGs (as it has been in other situations in the past).[36] The aforementioned independent report found that the EU not only needs to strengthen its leadership role in the achievement of the SDGs, but also "generates large, negative spillovers that impede other countries' ability to achieve the SDGs".

Although the views expressed in the 2019 ESDR report did not reflect the views of any organizations, agencies or programmes of the UN or the EU, it is worth mentioning that the report proposed six combined transformations that needed to be applied in the EU in order to achieve all 17 SDGs. It also offered practical recommendations for how the EU and its member states could achieve the SDGs, focusing on three broad areas: internal priorities, diplomacy and development cooperation, and negative international spill overs. On the plus side, the 2019 ESDR report recognized that instruments to achieve the SDGs already exist throughout the EU. Ultimately, the text recommended mainstreaming the SDGs.

Along the same lines, the 2019 Eurostat report also concluded that Europe was falling behind in terms of climate action and industry, innovation, and infrastructure, while progress was moderate or mixed on responsible consumption and production, life on land, reduced inequalities, affordable and clean energy, zero hunger, and gender equality. Moreover, the report noted that despite the EU's aim to eliminate its contributions to climate change by 2050, its plan on how to do so remains light on details.[37] By the same token, the 2018 EEA monitoring report on the 7th European Environmental Action Plan (EAP) has concluded that 23 out of the 30 targets of the 7th EAP are unlikely to be achieved by 2020, including the reduction in the food sector's environmental impact and the halting of biodiversity loss.

The EU's 2019 Annual Strategic Report on the implementation and delivery of the Sustainable Development Goals also recognizes that the EU "has the potential to deliver" a response. For this to happen, the report states that the European Commission and the member states must ensure a horizontal approach to the SDGs in their policies by "fully integrating the SDGs in EU policies and governance, providing guidance for both the EU institutions and the Member States in their implementation, monitoring and review of the 2030 Agenda, and outlining detailed roadmaps, concrete targets and deadlines". In other words, the SDGs are not simply an objective, but also a compass and a roadmap.[38] Without a holistic and integrated response, the SDGs will not be accomplished.[39] The question is how the EU will turn the SDGs into a reality.

4. The EU as a global trailblazer in sustainable development

The European scale is not expansive enough. An agenda that has a global impact is essential and "that agenda is best captured by the 17 Sustainable Development Goals".[40] These are the terms by which the EU intends to not only increase the scale of the EU, which is a global force to be reckoned with, but also to set an international agenda.[41] Despite all of its shortcomings, the EU is a role model and a trailblazer for the rest of the world, since it "has already put in place many of the most modern policies in the world to foster sustainability".[42] Indeed, the EU's intention to be a frontrunner in the implementation of SDGs is in line with the TEU, which provides that the EU is to "contribute to . . . the sustainable development of the Earth . . . as well as to the strict observance and the development of international law". The EU Commission highlights that the EU and the United Nations are natural partners in any efforts to tackle the SDGs, and that a rules-based system is the best guarantor for the sustainability of our economy and society. The EU Commission also underlines that this approach, together with multilateral diplomacy, can produce solutions to international challenges. In addition, the Global Strategy for the EU's Foreign and Security Policy determines that the SDGs should be a cross-cutting priority and that "concerted efforts are needed by the EU and its Member States in their dealings with the rest of the world".[43]

In the same vein, the European Parliament has called on the European Commission to strengthen its collaboration with the UN. In the Parliament's opinion, the Union should renew its commitment to being a global frontrunner and it specifies that this must be done in line with the principle of subsidiarity and in close cooperation with the EU's international partners. The European Parliament also holds that the EU's political engagement should be reflected in the multiannual financial framework (MFF) for 2021–2027, insisting that the 2030 Agenda should be an additional catalyst for a more coordinated approach between the EU's internal and external action and its other policies, and for greater coherence across Union financing instruments for a global response and commitment towards sustainable growth and development.[44] Ultimately, the EU's position as the world's largest single market, largest trader and investor, and largest provider of development assistance can have a major impact on the success of the United Nations' 2030 Agenda.

In this context, the EU's assertive new 2019 EGD[45] represents an attempt to respond to all of these situations and to the fact, among others, that "[o]ne million of the eight million species on the planet are at risk of being lost. Forests and oceans are being polluted and destroyed".[46] The Green Deal is a detailed mainstream policy document affecting every sector of one of the richest, most sophisticated economies in the world. For many, however, none of this will be enough if it fails to tackle profound structural transformations; in their opinion, it could merely result in a renewed attempt to perpetuate a specific liberal economic and social model.[47] The truth is, however, that the EGD presents important changes that are both specific and drastic. The EGD sets out "how to make Europe the first climate-neutral continent by 2050, boosting the economy, improving people's health and quality of life, caring for nature, and leaving no one behind".[48]

The EGD is also giving concrete shape to the EU's aim to have an impact on the whole world through the implementation of the SDGs. As recent EU trade negotiations show, the EU now insists that trade agreements with third countries must be used to drive environmental improvements.[49] The EGD recalls that "[t]he environmental ambition of the Green Deal will not be achieved by Europe acting alone".[50] Ultimately, the EGD seeks to make respect for the Paris Agreement an essential element of all future comprehensive trade agreements.[51] As another example of the EU's international role, the EGD also calls for a review of the rules on waste shipments. It even declares that, in its view, the EU should stop exporting its waste outside the EU.

5. From mainstreaming Sustainable Development Goals to the 2019 European Green Deal

For some, the idea of reconciling economic growth, social equity, and environmental protection (the so-called "triple bottom line") in a single concept as proposed by the Brundtland Commission has been a complete failure: We live now, "[f]our decades later, in a world scarred by inequality, looming climate catastrophe and the rupture of the Earth system".[52] However, the world should not measure the "success" of sustainable development mainly in terms of economic growth. Sustainable development may risk fostering the illusion that endless growth is feasible on a finite planet.[53] However, as Haydn Washington and Sam Alderman argue, "[i]n a finite world, we need to accept once and for all that sustainability cannot be about further growth".[54] This is absolutely one of the challenges that the EU faces.

In that spirit, all EU institutions need to take steps to ensure coordinated work in making progress towards the SDGs. Even if more drastic measures become necessary, a better EU regulation agenda can play an important role throughout the process. Officially launched in 2002, the EU's better regulation agenda is fully embedded today in the European Commission's policy process. In fact, the need to better combine regulation and sustainable development has been on the table since the beginning, as has the idea of mainstreaming environmental and other interests, such as social justice, in all policies. Without them, the SDGs cannot provide a fully actionable policy framework.[55]

Furthermore, when the new European Commission announced the EGD on 11 December 2019, it set out a wide range of major policy and legislative proposals to transition Europe to becoming "climate neutral" by 2050. Whether or not it is satisfactory, there is no doubt that the plan[56], which aims to "reset the Commission's commitment to tackling climate and environmental-related challenges that is this generation's defining task",[57] is specific. Unlike the US Democrats' Green New Deal,[58] the European Union's Green Deal is also technically feasible. It is concrete. As a result, it could do much more to pave the way for future environmental gains. In addition, the EGD is ambitious. It aims to decarbonize the world's second-largest economy within three decades. Indeed, for many, including the EU

Commission itself, the EGD represents the Commission's most ambitious environmental programme to date, seeking to affirm Europe as a global leader on the SDGs and climate change. As noted earlier and in the chapter on climate change, the EGD, if achieved, will have far-reaching consequences for everyone including EU member states, third countries, operators and investors across a wide range of sectors, and international exporters for whom Europe is a key market.

On the downside, the EGD is still only an "initial roadmap" of the key policies and measures needed to achieve the SDGs. Implementation will be crucial. For this reason, the EGD itself establishes that it will need to be updated as policy evolves. This might explain why the Communication is also light on details in relation to how the Commission proposes to deliver the EGD despite setting out an indicative timeline for implementation.[59] In any case, the EGD marks a dramatic shift, stating that the purpose is not only about mainstreaming environmental issues, but also about taking an entirely different approach that aims to guide all EU actions, regulations, and policies. It seeks to give a totally new direction to EU governance. Without exception, "all EU actions and policies will have to contribute to the European Green Deal objectives".[60] To this end, for example, the European Commission will develop a comprehensive plan by summer 2020 to increase the EU's 2030 greenhouse gas (GHG) emissions reduction target from 40% (compared with 1990 levels) to at least 50%, and towards 55% "in a responsible way", and it will pursue climate change legislation and related norms, which are addressed in detail in the chapter on climate change.[61]

Along the same lines, the EGD also notes that a more circular economy will necessitate changes to waste rules, including new legislation to tackle over-packaging and waste generation, as well as revisiting the rules on waste shipments and EU waste exports, which has been noted earlier, as well as in Chapter 4 on the promotion of sustainable development through multilateral trade.

As stated in other documents mentioned earlier in the chapter, such as the EU Reflection Paper "Towards a Sustainable Europe",[62] all of these efforts will require a "Sustainable Finance Strategy".[63] Ultimately, when the EGD directly mentions mainstreaming sustainable development in Section 2.2, it refers firstly to financial and budget aspects and secondly to research and education.[64]

Lastly, the EGD announces many other initiatives across a broad range of areas,[65] with each of these initiatives likely to require substantial legislative and policy changes in the coming decades. In this regard, the "evaluate first" principle dictates that the European Commission will first take stock of the problem that needs to be addressed and identify possible gaps and inconsistencies in legislation before proposing new rules. As for the challenges of implementation, the EU's 2002 better regulation agenda could be valuable.[66] Nowadays, the better regulation agenda is, as a matter of course, fully embedded in the European Commission's policy process. In a broad sense, better regulation implies that major EU policy initiatives should be backed with evidence that is collected and processed through a circular, learning-oriented approach called the "policy cycle" by the Commission.[67] The "policy cycle" also requires EU policies to be monitored over their lifetime and then periodically evaluated afterwards. The EGD maintains this requirement because, unfortunately, it would not be the first time that the lack of success in implementing sustainable development had to do with the fact that environmental agendas and their goals and premises were never mainstreamed efficiently into the policy process of the European Commission and other EU institutions. As Renda has stated in relation to past efforts:

> [W]hile at the highest political level EU institutions and member states were setting targets and ambitions for Europe's sustainable and inclusive growth, in their daily practice of policymaking EU institutions were dancing to a different tune and decisions were adopted on the basis of significantly different criteria and benchmarks.[68]

Against this background, the current use of better regulation in the European Commission, other EU institutions and member states do not appear conducive to mainstreaming sustainable development in daily regulatory practice. In this respect, however, Renda takes the "Commission's new commitment to better regulation and SDGs at face value", and "explore[s] the changes that might be needed to ensure that better regulation can be deployed in the most effective way to support the 2030 Agenda".[69] Today, with the EGD, the European Commission not only embraces the SDGs and better regulation, but has also established them as one of the pillars of the EGD.[70] Overall, the EGD appears to be an effective plan that could prove successful.

In addition, the EGD requires that "all EU actions and policies should pull together to help the EU achieve a successful and just transition towards a sustainable future". It also calls again on the better regulation tools ("the Commission's better regulation tools provide a solid basis for this" (section *2.2.5*, entitled "A green oath: 'do no harm'").[71] Certainly, better regulation tools imply public consultations; the identification of environmental, social, and economic impacts; and analyses to make efficient policy choices at a minimum cost. All of this is in line with the objectives of the EGD and with the SDGs. To achieve these aims, public participation is also key, and the EGD recognizes the role of stakeholders who are invited to use the available platforms to simplify legislation and identify problematic cases so that the Commission can consider their inputs when preparing evaluations, impact assessments, and legislative proposals for the European Green Deal.[72] Ultimately, without indicators and evaluations, there will be no coherence between current legislation and new priorities.

6. Final remarks

The EU is not only a role model for the rest of the world, but it has also become a frontrunner by already putting into place many of the most advanced and efficient policies to foster sustainability. It is also a natural partner of the UN in efforts to tackle the SDGs. Nevertheless, while Europe leads globally on the SDGs, none of the EU member states is on track to achieve the SDGs by 2030. Currently, only three are close to doing so: Denmark, Sweden, and Finland. Indeed, ecological sustainability is incompatible with current patterns of production and consumption.

Taking into account that SDGs are not merely an objective but also a compass and a roadmap, an integrated and holistic SDG strategy must be implemented. The EU has spearheaded sustainable development policies and made enormous strides towards implementing development that is sustainable, even with all the complications entailed in such an effort. To this end, previous EU tools, such as the better regulation agenda and the use of indicators, are valuable but not sufficient. Other existing instruments and mechanisms, such as budgets, investment strategies, regulatory governance, and monitoring frameworks, are also key to mainstreaming the SDGs, but they are also insufficient.

In that spirit, the EU's promising 2019 EGD could be a roadmap to success if it is fully implemented. Certainly, the EGD marks a dramatic shift, stating that the purpose is not only about mainstreaming environmental issues but also about taking an entirely different approach that aims to guide all EU actions, regulations, and policies. It seeks to give a totally new direction to EU governance. Without exception, "all EU actions and policies will have to contribute to the European Green Deal objectives".[73] For many, including the European Commission itself, the EGD represents the European Commission's most ambitious environmental programme to date, seeking to affirm Europe as a global leader on the SDGs and climate change. If achieved, the EGD will have far-reaching consequences for everyone. On the downside, it is still only an "initial roadmap" of the key policies and measures needed to achieve the SDGs.

Ultimately, in light of the global challenges, the evidence shows that without a structural shift and a holistic and integrated strategy to achieve the SDGs, such goals will not be accomplished. Europe is facing a global environmental emergency. Recent events show that the world, if necessary, is able to take drastic action when faced with an emergency. On the plus side, most experts agree that the EGD is feasible and that instruments now exist to achieve sustainable development that will not compromise future generations.

Notes

1 Rifkin, 2019, p. 1; Carson, 1962; Meadows et al., 2004.
2 World Commission on Environment and Development (WCED), 1987.
3 Sachs, 2015, p. 3; Williams and Millington, 2004, p. 101.
4 Adelman, 2018, pp. 15–40.
5 United Nations, *Stockholm Declaration of the United Nations Conference on the Human Environment*, UN Doc. A/Conf.48/14/Rev. 1(1973), Preamble.
6 Redclift and Woodgate, 2013, p. 92; Aguilera Vaqués, 2000.
7 Adelman, 2018, p. 6.
8 Adelman, 2013.
9 United Nations Conference on Environment and Development, *Rio Declaration on Environment and Development* (A/CONF.151/26, vol. I) and Agenda 21 (A/CONF.151/26, vol. II) 14 June 1992.
10 United Nations Conference on Sustainable Development, *Outcome Document: The Future We Want*, 20 to 22 June 2012. A/RES/66/288.

11 Adelman, 2013, p. 11.
12 UN General Assembly Resolution 70/1. *Transforming our world: the 2030 Agenda for Sustainable Development*, 25 September 2015, A/RES/70/1.
13 The SDGs (2015 to 2030), which were validated by all 193 member states, were part of a broader document called Transforming our world: the 2030 Agenda for Sustainable Development (approved unanimously by the General Assembly of the United Nations through Resolution 1/70).
14 ICSU/ISSC, 2015, p. 6.
15 EU Commission Communication: The European Green Deal, COM(2019)640 final, 11.12.2019.
16 Adelman, 2018.
17 Meadows et al., 2004; Rockström et al., 2009; Steffen et al., 2015; Sachs et al., 2019.
18 French and Kotzé, 2018; Barral, 2012.
19 Adelman, 2018, p. 17.
20 Fletcher and Rammelt, 2016.
21 European Economic and Social Committee (2018), *Opinion of the European Economic and Social Committee on 'Indicators better suited to evaluate the SDGs – the civil society contribution.* OJ No. C 440, 6.12.2018 p. 14.
22 Vajdaa and Rhimes, 2018.
23 Mariana Kotzeva, Director-General of Eurostat, Foreword, European Commission. Statistical Office of the European Union, 2019, p. 5; Eurostat, 2019, available from: https://ec.europa.eu/eurostat/documents/3217494/9940483/KS-02-19-165-EN-N.pdf/1965d8f5-4532-49f9-98ca-5334b0652820 [accessed 1 April 2020]: "This publication is the third of Eurostat's regular reports monitoring progress towards the SDGs in an EU context. The analysis in this publication builds on the EU SDG indicator set, developed in cooperation with a large number of stakeholders. The indicator set comprises around 100 indicators and is structured along the 17 SDGs. For each SDG, it focuses on aspects which are relevant from an EU perspective. The monitoring report provides a statistical presentation of trends relating to the SDGs in the EU over the past five years ('short-term') and, when sufficient data are available, over the past 15 years ('long-term'). The indicator trends are described on the basis of a set of specific quantitative rules". See also Papadimitriou et al., 2019.
24 Pallemaerts and Azmanova, 2006; Krämer, 2016.
25 Humphreys, 2017; Barral, 2012.
26 European Union Agency for Fundamental Rights, 2019, p. 6; Multi-Stakeholder Platform, 2018.
27 European Commission, Preparatory Acts. *Reflection Paper Towards a Sustainable Europe by 2030*, COM(2019)22, 30.1.2019.
28 European Economic and Social Committee (2018), *Opinion of the European Economic and Social Committee on 'Indicators better suited to evaluate the SDGs – the civil society contribution'*, OJ No. C 440, 6.12.2018 p. 14.
29 Von der Leyen, U., Political guidelines for the next Commission (2019–2024) – 'A Union that strives for more: My agenda for Europe' (online), available from: https://ec.europa.eu/commission/presscorner/detail/en/ip_19_6691 [accessed 5 February 2020].
30 COM(2019)640, p. 4.
31 The European Parliament Resolution of 14 March 2019, regrets that the European Commission has not yet developed an integrated and holistic SDG implementation strategy (P8_TA(2019)0220), European Parliament Resolution of 14 March 2019 on the Annual Strategic Report on the implementation and delivery of the Sustainable Development Goals (SDGs) (2018/2279(INI)), available from: www.europarl.europa.eu/doceo/document/A-8-2019-0160_EN.html [accessed 2 February 2020].
32 Hickel, 2015.

33 SDSN & IEEP, 2019.
34 The 2019 Europe Sustainable Development Report details SDG achievements by the EU and its member states. Based on a methodology that measures distance to object, the report assesses how fast each country – and the European Union as a whole – has been progressing towards each of the SDGs and evaluates whether the current pace of progress will be sufficient to achieve the targets by 2030. The Europe Sustainable Development Report is consistent with the EU's official Eurostat report, which also examines how far each country is from achieving the SDGs. The report builds on the methodology developed by the SDSN for the annual Sustainable Development Report, including the SDG Index and Dashboards, issued by the SDSN and Bertelsmann Stiftung to track countries on the SDGs since 2016. The findings of the 2019 ESDR independent report are consistent with major related publications.
35 See European Agency for Fundamental Rights, 2019; Fievet, 2001.
36 Gunder Frank, 2004.
37 See Chapter 2.
38 See President Juncker's declarations on 13 September 2017, in relation to the announcement of the Reflection Paper "Towards a Sustainable Europe by 2030, on the follow-up to the UN Sustainable Development Goals, including on the Paris Agreement on Climate Change", (2019), COM(2019)22.
39 EU Parliament Resolution of 14 March 2019 on the Annual Strategic Report on the implementation and delivery of the Sustainable Development Goals (SDGs) (2018/2279[INI]). In the same vein, this EU Parliament Resolution also affirmed that the four essential pillars of sustainable development (social, environmental, economic and governance) must be addressed in a comprehensive manner in order to achieve the Sustainable Development Goals (SDGs).
40 COM(2019)22 final.
41 Washington, 2015.
42 COM(2019)22 final.
43 *Ibid*. See also Chapter 4.
44 EU Parliament Resolution of 14 March 2019 on the Annual Strategic Report on the implementation and delivery of the Sustainable Development Goals (SDGs) (2018/2279[INI]). See also Chapter 4.
45 COM(2019)640 final.
46 *Ibid*. p. 2. Sources: (i) Intergovernmental Panel on Climate Change (IPCC): Special Report on the impacts of global warming of 1.5°C; (ii) Intergovernmental Science-Policy Platform on Biodiversity and Ecosystem Services: 2019 Global assessment report on biodiversity and ecosystem services; (iii) The International Resource Panel: Global Resources Outlook 2019: Natural Resources for the Future We Want; (iv) European Environment Agency: the European environment – state and outlook 2020: knowledge for transition to a sustainable Europe.
47 Adelman, 2018; Blühdorn, 2016; Washington, 2015; Hickel, 2015.
48 COM(2019)640 final, p. 5.
49 *Ibid*., pp. 20–22. See also Chapter 4.
50 COM(2019)640 final, p. 1, where the European Commission insists on "The EU can use its influence, expertise and financial resources to mobilise its neighbours and partners to join it on a sustainable path. The EU will continue to lead international efforts and wants to build alliances with the like-minded".
51 See Chapter 4.
52 Washington, 2015, p. 36.
53 Meadows et al., 2004.
54 Adelman, 2018; Washington, 2015, p. 36.
55 Renda, 2017, p. 7.
56 See Chapter 2.

57 COM(2019)640 final, p. 1.
58 Rifkin, 2019.
59 Available from: https://ec.europa.eu/info/files/annex-roadmap-and-key-actions_en [accessed 20 January 2020].
60 COM(2019)640 final, p. 3.
61 And, as happens with the SDGs, the EGD intends to cover all sectors of the economy, notably transport, energy, agriculture, buildings, and industries such as steel, cement, ICT, textiles, and chemicals. The Commission commits to launch many other strategies and plans for the next coming years too (e.g., the Biodiversity Strategy for 2030, the new Industrial Strategy and Circular Economy Action Plan, the Farm to Fork Strategy for sustainable food and proposals for a pollution-free Europe, etc.). For that purpose, for instance and among other measures, the EGD announces a Circular Economy Action Plan that promotes a sustainable products policy (to support the circular design of all products based on a common methodology), prioritize reducing and reusing materials before recycling them, and strengthen extended producer responsibility. While the mainstreaming of environmental interests has been at the centre of EU aims for decades, the EGD really goes into detail and determines that action will focus on resource-intensive sectors such as textiles, construction, electronics, and plastics. Moreover, the Commission announces that it will analyse the need for a "right to repair" and curb the built-in obsolescence of devices (in particular electronics). In the same vein, see Business and Sustainable Development Commission, 2017.
62 COM(2019)640 final.
63 For this purpose, the Commission, aware that delivering the EGD will require significant public and private investment, states that a renewed Sustainable Finance Strategy will be presented by autumn 2020 and added to the Sustainable Europe Investment Plan expected in early 2020. In the same vein, an EU taxonomy for environmentally sustainable investment will be adopted. However, the document also recognizes that, although a provisional agreement on the draft EU Taxonomy Regulation was reached on 5 December 2019, a final adoption by the Council and Parliament may be delayed until after 2020 due to disagreements over the treatment of nuclear energy. Moreover, the renewed Sustainable Finance Strategy is also expected to include announcements on an EU green bond standard and an EU ecolabel for retail investment products (both based on the forthcoming EU taxonomy), as well as proposals on how to better integrate climate and environmental risks into the financial system. The EGD Communication also explains that the Commission is proposing to review the Non-Financial Reporting Directive (Directive 2014/95/EU) in 2020.
64 Hence, "Mainstreaming sustainability in all EU policies" mainly refers to: "*2.2.1. Pursuing green finance and investment and ensuring a just transition*" and "*2.2.2. Greening national budgets and sending the right price signals*", followed by research and education: "*2.2.3. Mobilising research and fostering innovation*" and "*2.2.4. Activating education and training*", and finally tackling the transition towards a sustainable future: "*2.2.5. A green oath: 'do no harm'*", COM(2019)640 final, pp. 15–19.
65 These areas include, among others: a chemicals strategy for sustainability, by summer 2020; a zero pollution action plan for water, air, and soil, by 2021; a strategy on offshore wind, in 2020; an EU Biodiversity Strategy for 2030, by March 2020; a Farm to Fork Strategy, by spring 2020; and a strategy for sustainable and smart mobility, by 2020.
66 In the same vein, the 2015 European Commission Staff Working Document: Better Regulations for Innovation-Driven Investment, highlighted the positive impact of good regulation on innovation.
67 European Commission Communication: Smart regulation in the European Union, COM(2010)543, 8.10.2010.
68 Renda, 2017, p. 2.

69 *Ibid.*, p. 8.
70 Indeed: "In addition, building on the results of its recent stock taking of better regulation policy, the Commission will improve the way its better regulation guidelines and supporting tools address sustainability and innovation issues. The objective is to ensure that all EGD initiatives achieve their objectives in the most effective and least burdensome way and all other EU initiatives live up to a green oath to 'do no harm'. To this end, the explanatory memorandum accompanying all legislative proposals and delegated acts will include a specific section explaining how each initiative upholds this principle", COM(2019)640 final, p. 23.
71 COM(2019)640 final.
72 *Ibid.*, pp. 15–19. It also points out that: "The Commission will consider revising the Aarhus Regulation to improve access to administrative and judicial review at EU level for citizens and NGOs who have concerns about the legality of decisions with effects on the environment. The Commission will also take action to improve their access to justice before national courts in all Member States. The Commission will also promote action by the EU, its Member States and the international community to step up efforts against environmental crime" (p. 23).
73 COM(2019)640.

References

Adelman, S., 2013. Rio+20: Sustainable Injustice in a Time of Crises. *Journal of Human Rights and the Environment*, 4(1), 6–31.

Adelman, S., 2018. The Sustainable Development Goals, Anthropocentrism and Neoliberalism. In French, D. and Kotzé, L. (Editors), *Sustainable Development Goals: Law, Theory and Implementation*. Northampton: Edward Elgar Publishing, 15–40.

Aguilera Vaqués, M., 2000. *El desarrollo sostenible y la constitución española*. Barcelona: Atelier.

Barral, V., 2012. Sustainable Development in International Law: Nature and Operation of an Evolutive Legal Norm. *European Journal of International Law*, 23, 377–400.

Blühdorn, I., 2016. Sustainability -Post-sustainability-Unsustainability. In Gabrielson, T., Hall, C., Meyer, J.M. and Schlosberg, D. (Editors), *The Oxford Handbook of Environmental Political Theory*. Oxford: Oxford University Press, 259–273.

Business and Sustainable Development Commission, 2017. *Better Business Better World. The Report of the Business & Sustainable Development Commission*. London: Business and Sustainable Development Commission. Available from: https://sustainabledevelopment.un.org/content/documents/2399BetterBusinessBetterWorld.pdf.

Carson, R., 1962. *Silent Spring*. Boston. Houghton Mifflin Harcourt.

European Union Agency for Fundamental Rights, 2019. *Implementing the Sustainable Development Goals in the EU: A Matter of Human and Fundamental Rights*. Brussels: Publications Office of the European Union. Available from: https://fra.europa.eu/sites/default/files/fra_uploads/fra-2019-fundamental-rights-report-2019-focus_en.pdf.

Eurostat, 2019. *Sustainable Development in the European Union, Monitoring Report on Progress Towards the SDGs in an EU Context*. Brussels. Publications Office of the European Union. Available from: https://ec.europa.eu/eurostat/documents/3217494/9940483/KS-02-19-165-EN-N.pdf/1965d8f5-4532-49f9-98ca-5334b0652820.

Fievet, H., 2001. Réflexions sur le concept de développement durable: prétentions économiques, principes stratégiques et protection des droits fondamentaux. *Revue Belgue de Droit International*, 128–143.

Fletcher, R. and Rammelt, C., 2016. Decoupling: A Key Fantasy of the Post-2015 Sustainable Development Agenda. *Globalizations*, 2, 450–467.

French, D. and Kotzé, L. (Editors), 2018. *Sustainable Development Goals: Law, Theory and Implementation*. Northampton: Edward Elgar Publishing.

Gunder Frank, A., 2004. The Development of Underdevelopment. In Wheeler, S.M. and Beatley, T. (Editors), *The Sustainable Urban Development Reader*. London: Routledge, 38–41.

Hickel, J., 2015. Why the New Sustainable Development Goals Won't Make the World a Fairer Place. *The Conversation*. Available from: http://theconversation.com/why-the-new-sustainable-development-goals-wont-make-the-world-a-fairer-place-46374.

Humphreys, M., 2017. *Sustainable Development in the European Union: A General Principle*. London: Routledge.

ICSU/ISSC, 2015. *Review of the Sustainable Development Goals: The Science Perspective*. Paris: International Council for Science.

Krämer, L., 2016. *EU Environmental Law*. London: Sweet & Maxwell.

Meadows, D.H., Randers, J. and Meadows, D.L., 2004. *The Limits to Growth: The 30-year Update*. White River Junction: Chelsea Green Publishing.

Multi-Stakeholder Platform, 2018. *Europe Moving Towards a Sustainable Future: Contribution of the Multi-Stakeholder Platform on the Implementation of the Sustainable Goals in the EU, Reflection Paper*. Brussels. European Union.

Pallemaerts, M. and Azmanova, A., 2006. *The European Union and Sustainable Development: Internal and External Dimensions*. Brussels: BVUPRESS Brussels University Press.

Papadimitriou, E., Neves, A. and Becker, W., 2019. *JRC Statistical Audit of the Sustainable Development Goals Index and Dashboards*. Brussels: European Commission, Joint Research Centre.

Redclift, M. and Woodgate, G., 2013. Sustainable Development and Nature: The Social and the Material. *Sustainable Development*, 21(2), 92–100.

Renda, A., 2017. *How Can Sustainable Development Goals be 'mainstreamed' in the EU's Better Regulation Agenda?* Brussels: CEPS Policy Insights.

Rifkin, J., 2019. *The Green New Deal: Why the Fossil Fuel Civilization Will Collapse by 2028, and the Bold Economic Plan to Save Life on Earth*. New York: St. Martin's Publishing Group/Macmillan.

Rockström, J. et al., 2009. Planetary Boundaries: Exploring the Safe Operating Space for Humanity. *Ecology and Society*, 14, 2. Available from: www.ecologyandsociety.org/vol14/iss2/.

Sachs, J.D., 2015. *The Age of Sustainable Development*. New York: University of Columbia Press.

Sachs, J.D., Schmidt-Traub, G., Mazzucato, M., Messner, D., Nakicenovic, N. and Rockström, J., 2019. Six Transformations to Achieve the Sustainable Development Goals. *Nature Sustainability*, 2, 805–814.

SDSN & IEEP, 2019. *The 2019 Europe Sustainable Development Report. Sustainable Development Solutions Network and Institute for European Environmental Policy*. Paris and Brussels (ESDR 2019).

Steffen, W., Richardson, K., Rockstrom, J., Cornell, S. E., Fetzer, I., Bennett, E. M., et al., 2015. Planetary Boundaries: Guiding Human Development on a Changing Planet. *Science*, 347(6223), 730–735.

Vajdaa, C. and Rhimes, M., 2018. Greening the Law: The Reception of Environmental Law and its Enforcement in International Law and European Union Law.

Columbia Journal of European Law, 24 (Fall), 455–473. Available from: https://store. legal.thomsonreuters.com/law-products/Law-Reviews-and-Journals/Law-Reviews– Journals-Westlaw-PROtrade/p/104937407.

Washington, H., 2015. *Demystifying Sustainability: Towards Real Solutions*. London: Routledge.

Williams, C.C. and Millington, A.C., 2004. The Diverse and Contested Meanings of Sustainable Development. *The Geographical Journal*, 170(2), 99–104.

World Commission on Environment and Development (WCED), 1987. *Our Common Future*. Oxford: Oxford University Press.

10 Environmental justice in EU law and policies

A fundamental challenge[1]

Jordi Jaria-Manzano

1. Introductory remarks

Environmental justice has become a fundamental matrix of interpretation of the global environmental crisis. In reaction against the inequalities in the distribution of environmental harms in public policies,[2] the concept emerged in the United States (US) at the end of the 1970s, and gained visibility at a federal level with the mobilizations against the construction of a landfill in Warren County, North Carolina in 1982.[3] Indeed, the idea of environmental justice was not coined and developed in academic research or institutional discussions; rather, it evolved in the form of "grassroots responses to situations of environmental injustice, specifically the unequal and inequitable distribution of risks and benefits based on race and class".[4]

Given that the same kinds of problems were occurring all over the world, the idea of environmental justice eventually moved onto the global stage. In this context, environmental justice aims to confront the mass exploitation of natural resources within an institutional and legal framework that is essentially biased in favour of inequality and injustice.[5] It evolved into a general critique of the distribution of the outcomes of the global social metabolism, a conceptual framework coined in the context of ecological economics, conceived as the material exchange between society and nature that allowed the development of human societies through the use of natural resources.[6] The concept of social metabolism raises questions regarding the sustainability of human societies, and also regarding the justice of the distribution of resources and burdens in the process of exchange between nature and society, providing a conceptual basis for environmental justice discourses.[7]

Social metabolism acquires a planetary dimension in the context of the capitalist world-system, as long as capitalism is able to expand globally.[8] This expansion is ultimately the cause of the current global environmental crisis, paving the way for a geological transition – the passage from the Holocene to the Anthropocene.[9] However, though some interpretations tend to put the emphasis on humanity as whole as the cause of the geological change, it should be underlined that the human species acts as a geological agent only when organized in a singular form, i.e., the capitalist world economy, structurally divided between centre and periphery. The transformation of the planet should to be attributed fundamentally to the centre, as well as the benefits obtained from it.[10]

This duality, possibly complemented by a semi-periphery, is built upon relations of unequal exchange, resulting in an unfair distribution of the outcomes of global social metabolism.[11] The growing expansion of global social metabolism is causing both a crisis of sustainability and a situation of profound injustice, in so far as "the drivers of inequality and the drivers of the Anthropocene are largely the same".[12] For this reason, environmental justice takes on a global dimension, demanding a particular responsibility from the institutional actors in the centre of the capitalist word economy, such as the European Union (EU).[13]

This is the context in which environmental justice, a concept coined in the aftermath of civil rights struggles in the US, acquires a central role in the reaction of the Global South to the global environmental crisis and the geological transition. Thus, the idea of environmental justice takes into account the scarcity and vulnerability of natural resources, the social distribution of environmental liabilities and assets, and, finally, the diversity of human communities.[14] For this reason, though "there is no universally agreed upon definition of environmental justice, all definitions converge upon the concepts of discrimination and distribution of harms/risks and benefits".[15] Furthermore, I think that environmental justice should not be limited to distributional issues, but should also take into account recognition, evolving into a complex concept aiming to remedy the dispossession and the neglect of less-favoured populations.[16] This need for recognition is related to the empowerment of local communities and the respect for plurality.[17]

Consequently, environmental justice consists in the quest for a more equitable distribution of the outcomes of social metabolism, recognizing the various perspectives available regarding how to value these outcomes.[18] We are, therefore, faced with a potential constitutional idea, dealing with the limitation of available resources, as well as the need to set rules of distribution.[19] The fight against exclusion and the protection of the environment unfold together.[20] In view of the EU's growing leadership in establishing a framework for environmental protection, Antypas et al. maintain that "the time to confront the problem of environmental injustice in Europe has arrived".[21] In any case, the idea of environmental justice used here is more all-encompassing than the one contained in the Aarhus Convention, centred on the access to the courts, which has been incorporated to some extent into EU law given that the Union has been part of the Convention since 2005.[22] In the following pages, I address the question of how environmental justice, in a broad sense, can be embedded in EU law and policies.

2. Environmental justice vs sustainable development: conceptual considerations about the interpretation of the global environmental crisis

In the domain of law, environmental justice is the core concept in a critical approach to current legal solutions, which mostly endorse a managerial approach to the global environmental crisis and ignore the injustices embedded in the hegemonic processes of social reproduction.[23] For this reason, environmental justice is a potential driver of legal order – even if only in a limited way, as in the case of the US, where environmental justice is designed to confront environmental racism, focusing on the damage suffered by lower-income communities and on the ineffectiveness of public policies in addressing the deficits that occur in relation to the distribution of environmental liabilities.[24] Moreover, according to this approach, environmental justice also demands to improve citizen participation in decision-making, since "low-income communities and communities of color face a number of obstacles in . . . participation".[25]

In the end, however, this is a very limited idea, given that it focuses exclusively on the distribution of environmental burdens and eludes making an overall assessment of the global social metabolism and its unfair outcomes in the context of late capitalism.[26] It seems reasonable to extend the idea of environmental justice to the set of relations between society and nature, considering both the benefits and the damage derived from the functioning of the social metabolism, and thus confronting the relations of unequal exchange in its integrity. This is the orientation preferred in the construction of the concept in the Global South, expressed through concepts as significant as ecological debt.[27]

Therefore, environmental justice should be conceived as a comprehensive framework addressed to reversing the significant injustices in the global social metabolism. These injustices affect, above all, the Global South, but also the poor communities in rich countries, disadvantaged in terms of both their exposure to environmental hazards and of their restricted opportunities to enjoy the social gains resulting from the use of natural resources. Consequently, in my opinion, the idea of environmental justice emerges as a fundamental alternative to mainstream policies and regulations for confronting the global environmental crisis which are driven by the notion of sustainable development. Accordingly, environmental justice aims to challenge the mechanisms of unequal exchange and the hegemonic strategies of social reproduction based on the externalization of environmental costs, and to limit their repercussion on the local communities in the periphery.[28]

The concept of sustainable development is not intended to have a significant impact on the institutional structure of global capitalism, or to reverse the regulatory frameworks regarding trade and investment that drive the process of capitalist accumulation.[29] Defined in the World Commission on Environment and Development report entitled *Our Common Future*, "sustainable development" was characterized as such if it "meets the needs of the present without compromising the ability of future generations to meet their own needs".[30] Later, the 1992 Rio Declaration on the environment and development[31] defined sustainable development as its key concept – the core of the international community's response to the global environmental crisis. Since then, it has been vigorously embraced in a vast array of documents and regulations, at international and national level, and has become an "overarching societal objective".[32]

Soon after the Rio Conference, the EU went some way towards incorporating the concept of sustainable development in the Treaty on the European Union (TEU), signed in Maastricht on 7 February 1992. There are some implicit references to sustainable development in the Treaty establishing the European Economic Community (TEEC) (in article B, or amendment of article 2), and some explicit ones as well (in new article 130).[33] The concept is mentioned in various sections of the TEU, such as the preamble (the parties are "determined to promote economic and social progress for their peoples, taking into account the principle of sustainable development"), sections 2 (the Union "shall work for the sustainable development of Europe") and 5 (the Union "shall contribute to peace, security, the sustainable development of the Earth") of article 3; and sections d (the Union shall "foster the sustainable economic, social and environmental development of developing countries, with the primary aim of eradicating poverty") and f (the Union shall "help develop international measures to preserve and improve the quality of the environment and the sustainable management of global natural resources. In order to ensure sustainable development") of article 21.2.[34] Sustainable development is thus engrained in EU fundamental law and defines the foundations of EU environmental policies.

Mainstreamed since the early 1990s, sustainable development has become consolidated as a notion providing a conceptual framework for developing environmental policies inside the capitalist world economy, maintaining the socio-ecological trends of late capitalism. This is clearly reflected in the most recent interpretation of its implications in the Resolution adopted by the United Nations (UN) General Assembly as an outcome of the United Nations Conference on Sustainable Development held in Rio de Janeiro in June 2012.[35] This document endorses a developmental approach and a managerial orientation in defining the response of the international community to the environmental crisis, complementing the notion of sustainable development with the idea of the green economy in the aftermath of the financial crisis of 2007.[36]

In the end, then, the adoption of the concept of sustainable development helps to legitimize the persistence of the capitalist world economy as a hegemonic model of social reproduction, and endorses the institutional mechanisms for outsourcing environmental costs based on unequal exchange.[37] For this reason, sustainable development is, in principle, a notion opposed to environmental justice and its broad critique (at the international level at least) of the rules and outcomes of the social metabolism of global capitalism. Certain exceptions to this general rule should be mentioned, particularly in reference to Sustainable Development Goals (SDGs), as I will highlight later in this chapter, but basically there is a fundamental opposition between these two basic notions.

In any case, sustainable development pervades EU environmental policies and law. Therefore, in principle, any attempt to introduce the idea of environmental justice in the context of EU law seems doomed to failure. However, given its commitment to the correct and just management of scarce resources at the global level, environmental justice is closely linked to sustainability and social justice, which are also part of the axiological foundations of EU law and are firmly rooted in human rights.[38] In fact, the EU has, to some extent, given way to environmental justice, albeit in a narrow sense, restricting its possibilities to the procedural ideal of "securing standing for individuals and civil society organisations in environmental proceedings".[39] Accepting the opposition between environmental justice in a broad sense and sustainable development, I will explore current EU law to find possible ways in which environmental justice might be incorporated into EU law and policies in order to increase its commitment to sustainability and justice.

3. Environmental justice and fundamental EU values

EU law is founded upon the idea of human dignity, which is the first value mentioned in article 2 of the TEU. Human dignity is related to the recognition of fundamental rights. Accordingly, article 2 establishes that "[t]he Union is founded on the values of . . . respect for human rights, including the rights of persons belonging to minorities". The protection of human dignity and human rights is to be secured in an equitable way. This is underlined in this same article, which refers to pluralism, non-discrimination, justice, and solidarity – all values that have been associated with environmental justice by social movements over decades.[40] This framework seems to be related to the specific Western European tradition of the social state with the belief that the conditions in which people live are essential to an effective enjoyment of human rights.[41]

Human rights have acquired a fundamental position in the EU law primarily through the case law of the European Union Court of Justice (EUCJ).[42] They were mentioned relatively early in Stauder, a decision which gave a constitutional value to fundamental rights in EU law.[43] The Court developed this idea in cases such as Internationale Handelsgesellschaft[44] and most famously Nold,[45] which paved the way for establishing EU standards in this area. It also influenced later developments in the treaties which, to an extent, can be said to culminate in the Charter of Fundamental Rights of the European Union, which starts by establishing human dignity as the basis for the subsequent rights recognized in the document.[46]

Moreover, the possible accession of the EU to the European Convention on Human Rights (ECHR)[47] would have an important bearing on environmental protection, and might also be significant from the point of view of environmental justice.[48] The accession is still under negotiation, but it seems clear that the ECHR might exert a real influence in the interpretation of human rights in the EU, given that, to some extent, it shapes the shared values of the member states.[49] However, it should be stressed that this accession would not modify the competences of the Union (article 6.2 TEU), and so would add nothing to remedy the member states' shortcomings in this area.

The right to a healthy environment has naturally been developed as part of the mainstream response to the environmental crisis, gravitating around the fundamental notion of sustainable development, and being considered as a human right *in statu nascendi*.[50] Specifically, this right has been explored by the European Court of Human Rights, even though the Convention does not refer explicitly to it.[51] However, the relationship between human rights and environmental justice is complex. The mainstream narrative of human rights has historically tended to promote the creation of an individual space of autonomy, which ultimately reinforces the original bourgeois *ethos* of the first liberal rights and promotes the appropriation and exploitation of nature.[52] In fact, the European social state has been built on a massive exploitation of natural resources, deployed in the context of unequal exchange relations between the centre and the periphery, and causing a constant growth of global social metabolism.[53]

This situation has accelerated the dynamics of unsustainability and inequality that give rise to the environmental crisis and, ultimately, to the transition to the Anthropocene. Therefore, it is not surprising that the protection of the environment maintains a certain tension with the growing recognition of rights in the context of the evolution of constitutionalism.[54] Ultimately, we are facing an internal contradiction, to the extent that the growing exploitation of the Earth System to generate well-being produces, in reality, deficits in sustainability and equality that threaten the well-being that (it is claimed) are being pursued.[55] For this reason, the link between human rights and environmental justice, although significant, is problematic.

Perhaps this tension can be resolved through an alternative reading of the meaning of human rights. From positions anchored in the discourse of environmental justice, proposals have been made to move towards a transformation of the rights' paradigm in accordance with a revaluation of responsibility required by the global consequences of the transformation process that the Anthropocene implies. In this interpretation, human rights are configured as "critical concepts of resistance";[56] so they can be interpreted as patterns of protection for the most vulnerable and, in this sense, be linked to the ideas of responsibility and care.[57] Thus, an ecological existential minimum can be created beyond the individualistic tradition that is also projected in the European constitutional tradition of the social state.[58]

Indeed, it might be possible to read the treaties in this light. Article 3.1 of the TEU states that its aim is the well-being of the peoples belonging to the Union, and article 2 of the TEU refers to pluralism, non-discrimination, tolerance, justice, solidarity, and equality between women and men. Even accepting the individualistic roots and limitations of the social state in the context of a global ecological crisis (or rather, of geological transition), it seems possible to put forward an interpretation of the values of the EU – the centrality of human dignity and the attachment to human rights in a way that brings them closer to environmental justice – attributing particular relevance to the idea of well-being consecrated in the TEU.[59] The ambiguity of fundamental rights, encompassing in EU law "not only the classic guarantees of freedom from interference by the state, but also the right to participation and certain aspects of political, social and economic life",[60] allows a reading that is more in line with the idea of environmental justice. Moreover, the non-discrimination clause, developed in the secondary legislation of the Union, remains relevant.[61]

In this way, a new idea of well-being, based on the concept of quality of life and linked to responsibility and participation, can be explored.[62] It would imply a moderate degree of individual self-determination, at the same time as the assumption of responsibilities in relation to others, to life and to the Earth System as a whole, in a way that to some extent conforms to the theory of capabilities developed by Amartya Sen.[63] According to this perspective, despite their current weakness, environmental rights may represent a valid way to boost the demand for responsibilities in relation to the deterioration of the environment.[64] Accordingly, rights are conceived as a way of protecting embodied vulnerability rather than as a vehicle for developing individual self-determination.[65]

This suggests the need to reformulate the idea of a reasonable standard of living in accordance with the limitations of the Earth System and the responsibilities entailed by the advent of the Anthropocene.[66] It should be understood that the purpose of distributive justice is to compensate for the unwanted effects of the deployment of commutative justice in traditional constitutionalism, thus providing an opportunity for all members of the global community to enjoy the minimum conditions of life by reducing inequality in the distribution of resources.[67] In fact, this has been developed in a particular manifestation of environmental justice – climate justice – in which human rights have supported claims in the international arena.[68]

Climate justice addresses the shortcomings of the global social metabolism from the point of view of climate change, underlining how it develops into extreme inequalities which affect the living standards of different communities in the Global South.[69] It is based on the idea that the dignity of the person cannot be fully achieved when climate conditions impede the development of a healthy human life. For this reason, climate justice establishes a link between climate change – the most conspicuous manifestation of the Anthropocene narrative – and human rights, "to achieve a human-centred approach, safeguarding the rights of the most vulnerable people and sharing the burdens and benefits of climate change and its impacts equitably and fairly".[70]

This idea can be extended to environmental justice as an all-encompassing concept to confront the multiple injustices in the global social metabolism. It might represent a promising way of reconstructing human rights – and, therefore, fundamental rights in the EU – in greater harmony with environmental justice. In a context of anthropogenic planetary transformation driven by the global social metabolism of the capitalist world economy, this allows an interpretation of human rights aligned with environmental justice, particularly in the context of EU law.[71]

However, this possibility cannot hide the fact that EU law is strongly linked to the hegemonic idea of development. Even its treaties endorse a managerial response to the environmental crisis and the geological transition along the lines of *The Future We Want*. The question is how to overcome the mainstream strategies for confronting the geological transition and propose responses that are more sensitive in terms of justice and sustainability regarding the internal distribution of the outcomes of the social metabolism and the impact of the EU in global material flows, reducing its ecological footprint and facing up to its ecological debt.

4. EU law, EU policies, and environmental justice: vulnerability, conflicts, and change

The predominance of business-as-usual responses to the global environmental crisis is linked to the incapacity of the international community to find a sufficiently broad consensus in relation to a comprehensive reaction beyond the hegemonic forms of social reproduction. Consequently, international law seems to be incapable of offering anything else than slight corrections to the capitalist world economy. In fact, the very dynamics of the process of capitalist accumulation result in a form of global regulatory capture and translates into normative complexes that favour the mass exploitation of resources, aggravating the consequences of anthropic action on the Earth System.[72] Despite the possibilities offered by the EU legal order in terms of enhancing environmental justice, the achievement of significant changes through the conventional decision-making processes in the Union is not to be expected any time soon, at least through regulatory procedures.[73]

However, there are some strategies that can be implemented. Basically they are to do with a justice-oriented reading of sustainable development in the EU's fundamental strategies. In accordance with UN's SDGs: for example, through the European Green Deal.[74] Moreover, the directives implementing the Aarhus Convention could be linked with EU anti-discrimination frameworks to enhance the sensitiveness of EU law to environmental justice issues.[75] This possibility seems to depend heavily on the judicial application of EU law by national courts as well as by the EUCJ.

Indeed, in a context of regulatory capture, the construction of alternative legal narratives seems to find a way through the judicial resolution of conflicts; in this case, socio-environmental conflicts. Given the inability of institutional actors to move towards inclusive, comprehensive, and systematic responses to the global environmental crisis, movements articulated from civil society have been forcing legal innovation through conflict and, particularly, litigation.[76] The development of climate litigation in this respect is a remarkable phenomenon and, in fact, has become a powerful instrument available to civil society in its attempts to influence climate policies.[77] Cases such as Urgenda (2019) in the Netherlands,[78] linked directly to climate policies, and Generaciones Futuras (2018) in Colombia[79] point to significant changes in this regard.

From this perspective, climate justice – and environmental justice in general – is closely aligned with minority strategies against a mainstream politics that supports the hierarchies and divides of the capitalist world economy. For this reason, social activism is expected to increase the impact of environmental justice in EU law, particularly by bringing environmental conflicts into the courts. There, new concepts for transforming policies and regulations can be developed more easily than in the field of institutional decision-making processes. In fact, "a significant shift in power in favor of the judiciary when it now assumes an important judicial review function to keep executive and legislative in check" has been detected.[80]

To the extent that litigation will be based on the EU norms relating more to environmental justice, as discussed in the previous section, transformations in this field are to be expected, even if they are only fragmentary.[81] From this perspective, the creation of social consensus associated with the production of legislation should be constructed in a provisional form, responding to specific social conflicts, recognizing that legal discourse as such is conflictive and demands a profound sense of social and political integration in judicial practices.[82] In this scenario, the role of the EUCJ as a constitutional court is crucial. In particular, in protecting individuals from violations of fundamental rights[83] it is not unreasonable to imagine the EUCJ playing this role with regard to the internal policies and regulations of the Union, and also influencing the Union's behaviour on the international scene given its place inside the world economy.

5. Final remarks

EU law does not contain an explicit recognition of environmental justice as a fundamental principle in the development of its environmental policies and its regulations. Moreover, in principle, the importance of sustainable development as a central notion in EU environmental law rather suggests that environmental justice is paid little attention at present. However, the importance of human rights as a transversal element in the constitutional definition of the Union, as well as a re-reading of these rights in inclusive terms focusing on the protection of vulnerabilities rather than on the creation of spaces for individual self-determination, allows us to imagine a reinterpretation of EU law in terms that are more favourable to the achievement of environmental justice.

However, there is little reason to expect any active endorsement of environmental justice in the EU regulations. Rather, it seems that the formalization of environmental conflicts through litigation is the way to generate the momentum needed to enhance environmental justice concerns, both in the internal sphere of the EU and with regard to its international involvement. The role of the courts, particularly the EUCJ, is crucial in this scenario. Consequently, social movements take on a fundamental importance in the advancement of environmental justice in the EU. This social engagement can help to remodel the hegemonic concept of sustainable development and to explore more equitable solutions in the context of geological transition.

Notes

1 This work has been carried out in the context of the project "Global Climate Constitution: Governance and law in a complex context", funded by the Spanish Ministry of Economy and Competitiveness for the three-year period 2017–2019 (DER2016–80011-P; main researchers: Jordi Jaria-Manzano and Susana Borràs Pentinat).
2 On the inequalities linked to the environmental crisis, see Malm and Hornborg, 2014; Bonneuil and Fressoz, 2017, p. 65ff. As pointed out by Falk, 2009, p. 40, "[t]he tendency of environmentalist is to focus on their sense of what is causing the problems, and offer prescriptions designated to mitigate or end the perceived threat. This inattention to justice perspective tends to benefit the rich and powerful, as well as those currently alive, and to accentuate the burdens and grievances of the poor and marginal, and the unborn".
3 Gauna, 1995, p. 9.
4 Mickelson, 2009, p. 300; See also McWilliams, 1994, p. 761ff.
5 Anand, 2004, p. 15; Jaria-Manzano, 2012, p. 18.
6 Regarding the central role of the idea of social metabolism in ecological economics, see Fischer-Kowalski, 1998; Fischer-Kowalski and Hüttler, 1998; Weisz. 2007.
7 Jaria-Manzano, 2019, pp. 31–32.
8 This approach was introduced and developed by Wallerstein, 1974.

9 The idea of a new geological era was circulated by Crutzen and Stoermer, 2000; Crutzen, 2002. Since then, it has been capable to help the developing of Earth System Science as well as to impact in the geology, specially, after geologist Jan Zalaslewicz proposed in 2005 to test using formal geological criteria the effective occurrence of a geological change. The developments in Earth System Science have taken for granted the idea of geological transition, while it is still matter of discussion in geology. See for all this evolution, see Castree, 2019. In any case, on 21 May 2019, following guidance from the Subcommission on Quaternary Stratigraphy and the International Commission on Stratigraphy, the Anthropocene Working Group completed a binding vote accepting that the Anthropocene should be treated as a formal chrono-stratigraphic unit. See Anthropocene Working Group, 2019.

10 The considerations about justice and the importance of allocating particular responsibilities to some groups within the human species seems to be absent in the mainstream Anthropocene literature, as shown in Crutzen, 2002; Steffen et al., 2007. This has been challenged in social sciences literature, pointing out to justice issues regarding the geological transition. See Baskin, 2019, pp. 155–156. Here the distribution of resources and technologies in the world economy, along the divide between centre and periphery, is relevant. The differentiation between centre and periphery in the global economy was proposed by the Argentine economist Raúl Prebisch, working for the Economic Commission for Latin America and the Caribbean (ECLAC) in 1949. Since then, it has been used for many theorists to explain the structure of the world economy, particularly for Wallerstein. For an overview about this distinction, see Taylor and Flint, 2011, p. 20. In short, the Anthropocene is the "Age of *Some* Humans" rather than the "Age of Humans". See Wapner, 2019, p. 218.

11 Hornborg, 2009.

12 Baskin, 2019, p. 160.

13 Burke and Fishel, 2019, p. 102.

14 Jaria-Manzano, 2012, p. 20.

15 Antypas et al., 2009, p. 8.

16 Figueroa, 2011, p. 235.

17 Schlosberg, 2004, p. 537.

18 Hiskes, 2009, p. 25.

19 Jaria-Manzano and Borràs, 2019, p. 3. This idea is not fully strange in the context of international law, although it has been not properly developed and mostly ignored. For example, the Havana Charter for an International Trade Organization, which was signed in the Cuban capital in 1948 but never entered into force because of lack of enough ratifications, referred to "an equitable share of the international supply" of "products in . . . short supply" (article 45.1.b.i), addressing the unequal distribution of some commodities in the international markets. The Charter is available from: www.wto.org/english/docs_e/legal_e/havana_e.pdf [accessed 29 March 2020]. This idea has been transferred to the current text of the General Agreement on Tariffs and Trade (article XX.j), without significant effect in the regulation of international trade. I owe this reference to Professor Xavier Fernández Pons.

20 The relationship between poverty eradication, pursuit of a more equitable distribution of resources, and protection of the environment has been defended almost since the beginning of the environmental crisis. Thus, during the Stockholm Conference in 1972, there was a broad consensus on the fact that poverty and social instability made a sustainable system impossible. Therefore, the ideal of overcoming poverty and the realization of social justice should be linked with the possibility of preserving the environment under appropriate conditions. In this regard, see Jositsch, 1997, p. 99.

21 Antypas et al., 2009, p. 8.

22 See the Convention on access to information, public participation in decision-making, and access to justice in environmental matters, done at Aarhus, Denmark, on 25 June 1998, UNTS, vol. 2161, p. 447. See also the Council Decision of 17 February 2005

on the conclusion, on behalf of the European Community, of the Convention on access to information, public participation in decision-making, and access to justice in environmental matters (OJ No. L 124, 17.5.2005). The accession of the EU to the Convention has had as a result the Directive 2003/4/EC of the European Parliament and of the Council of 28 January 2003 on public access to environmental information and repealing Council Directive 90/313/EEC, OJ No. L 41, 14.2.2003, and the Directive 2003/35/EC of the European Parliament and of the Council of 26 May 2003 providing for public participation in respect of the drawing up of certain plans and programmes relating to the environment and amending with regard to public participation and access to justice Council Directives 85/337/EEC and 96/61/EC, OJ No. L 156, 25.6.2003; and the Regulation (EC) No 1367/2006 of the European Parliament and of the Council on the application of the provisions of the Aarhus Convention on Access to Information, Public Participation in Decision-making and Access to Justice in Environmental Matters to Community institutions and bodies, OJ No. L 264, 25.9.2006. Nevertheless, the importance of procedural rights regarding environmental justice should to be underlined. See, particularly, Antypas et al., 2009, p. 18. Moreover, some provisions in the Convention have some potential from the point of view of environmental justice in the sense of the expression that I use in this paper. For example, article 3.9 of the Conventions forbids any form of discrimination due "to citizenship, nationality or domicile".

23 Baskin, 2019, p. 160.

24 The Environmental Protection Agency (EPA) has defined environmental justice as "the fair treatment and meaningful involvement of all people regardless of race, color, national origin, or income, with respect to the development, implementation, and enforcement of environmental laws, regulations, and policies". This should imply "the same degree of protection from environmental and health hazards" and "equal access to the decision-making process to have a healthy environment in which to live, learn, and work". See United States Environmental Protection Agency, n.d.

25 Foster, 1999, p. 186.

26 Cole and Foster, 2001, p. 66; Cutter, 1995, p. 112.

27 As environmental justice, ecological debt is a notion developed by social movements. In this case, at the international level, it was circulated by activists in the Global Forum that was held parallel to the 1992 UN Conference on Environment and Development as a tool to compensate historical liabilities from the North toward the South and promote global justice. See Mickelson, 2007, pp. 274–275. Inspired by the concept of financial debt, ecological debt is presented as a claim to calculate and compensate the differential in favour of the system's periphery in the accounting of its exchange with the system's centre, based on the past and ongoing plundering of natural resources and traditional know-how, and use of the share of commons corresponding to peripheral countries, according to the definition of the Southern Peoples Ecological Debt Creditors Alliance (SPEDCA). See Jaria-Manzano et al., 2016, p. 383; Azar and Holmberg, 1995; Martinez-Alier, 2002; Paredis et al., 2004.

28 Anand, 2004, p. 139.

29 Dobson, 1998, p. 60. However, there have been optimistic readings of the notion, endorsing management strategies in relation to the global environmental crisis. See, particularly, Dernbach and Cheever, 2015.

30 World Commission on Environment and Development, 1987, p. 41. In any case, this seminal document did not appear out of nowhere, but rather responded to ideas that were already in circulation at the time its publication. See Walters, 1991, p. 424. For an overview and basic criticism on sustainable development, see Alder and Wilkinson, 1999, p. 127ff; Clarkson and Wood, 2009, p. 124ff).

31 *Report of the United Nations Conference on Environment and Development*, UN Doc. A/CONF.151/26 (Vol. I), 12 August 1992, Annex I.

32 Ebbeson, 2009, p. 1.

33 Treaty on European Union (original version). Available from: https://europa.eu/european-union/sites/europaeu/files/docs/body/treaty_on_european_union_en.pdf [accessed 23 January 2020].
34 Consolidated Version of the Treaty on European Union [2012] OJ C326/13.
35 General Assembly Resolution 66/288 [2012] The future we want.
36 See González, 2015, pp. 168–169.
37 Girardi, 2014, pp. 125–126.
38 Lazarus, 1994.
39 Antypas et al., 2009, p. 9.
40 On the idea of human dignity related to environmental justice, Mesa Cuadros, 2011, p. 31.
41 Wildhaber, 1972, p. 373.
42 Tridimas, 1999, p. 202.
43 Case C-29/69, Erich Stauder v City of Ulm, 12.11.1969. ECLI:EU:C:1969:57. Sustaining its constitutional dimension, see Cienfuegos Mateo, 2019, p. 4.
44 Case C-11/70, Internationale Handelsgesellschaft mbH v Einfuhr- und Vorratsstelle für Getreide und Futtermittel, 17.12.1970. ECLI:EU:C:1970:114.
45 Case C-4/73, J. Nold, Kohlen- und Baustoffgroßhandlung v Commission of the European Communities, 14.05.1974. ECLI:EU:C:1974:51.
46 OJ No. C 326/1, 26.10.2012.
47 Treaty Office on http://conventions.coe.int [accessed 20 April 2019].
48 Gaja, 2012.
49 The accession of the EU to the ECHR is a long-debated issue, which received a decisive impulse with the Treaty of Lisbon. However, the Opinion 2/13, European Convention on Human Rights, 18.12.2014. EU:C:2014:2454, introduced significant nuances in the implications of the accession. After half a decade since the decision, the institutions of the EU have addressed the issues pointed out by the EUCJ. Recently, the Commission has informed the Council of Europe that it has addressed the objections of the Court and it is ready to resume the negotiations.
50 Bosselmann, 2008, p. 12.
51 See, particularly, Powell and Rayner v United Kingdom (1990) 12 EHRR 355, López Ostra v Spain (1994) 20 EHRR 277, and Guerra and others v Italy (1998) 26 EHRR 357.
52 Escribano Collado, 1991, p. 3731.
53 Gerlach, 1989, p. 34.
54 Alexy, 2003, p. 37.
55 Canosa Usera, 1996, p. 81; Serrano Moreno, 1992, p. 52.
56 Grear, 2010, p. 39.
57 Marquet Sardà, 2010, p. 80.
58 Murswiek, 1995, p. 47.
59 For a reading of the well-being in line with the protection of the environment, see Untawale, 1990, p. 372.
60 Anderson and Murphy, 2012, p. 161.
61 See, particularly, Council Directive 2000/43/EC of 29 June 2000 implementing the principle of equal treatment between persons irrespective of racial or ethnic origin, OJ No. L 180, 19.7.2000.
62 Jositsch, 1997, p. 97; Murswiek, 1995, p. 50.
63 Cortina, 2002, p. 203ff; Thürer, 2001, p. 59.
64 Ewald, 2011, p. 416.
65 Westaway, 2012, p. 68; Atapattu, 2019, p. 200ff.
66 Karpen, 1988, p. 21.
67 Llano, 1988. p. 188.
68 Derman, 2019, p. 340ff.

69 On the relationship between climate justice and environmental justice, Brunnée, 2009.
70 Mary Robinson Foundation, n.d.
71 Accordingly, a reading of the United Nations' Sustainable Development Goals (SDGs) in line with environmental justice has been proposed by Gellers and Cheatham, 2019. See General Assembly Resolution 70/1 [2015] Transforming our world: the 2030 Agenda for Sustainable Development.
72 On regulatory capture, see Dal Bó, 2006. This can be particularly appreciated in climate policy. On the incapacity of institutional arrangements in providing effective and equitable responses to climate change, see Adelman, 2010, p. 163.
73 Nogueira López, 2011, p. 134.
74 European Commission Communication: The European Green Deal, COM(2019)640 final, 11.12.2019.
75 Antypas et al., 2009, p. 21.
76 This was noted by Allen and Lord, 2004, p. 552, more than a decade ago.
77 Kjellén, 2009, p. 334.
78 On 20 December 2019, the Dutch Supreme Court, the highest court in the Netherlands, upheld the previous decisions in the Urgenda Climate Case and confirmed the obligations of the Dutch government regarding an urgent and significant reduction of emissions of GHGs, which is linked to its human rights obligations. An unofficial English translation of the decision is available from: www.urgenda.nl/wp-content/uploads/ENG-Dutch-Supreme-Court-Urgenda-v-Netherlands-20-12-2019.pdf [accessed 5 February 2020].
79 On 5 April 2018, the Colombia Supreme Court reversed the decision of a lower court, recognizing that the "fundamental rights of life, health, the minimum subsistence, freedom, and human dignity are substantially linked and determined by the environment and the ecosystem". Moreover, the Constitutional Court recognized entitlement of the Amazon to some constitutional rights, commanding the government to address deforestation. An unofficial English translation of some excerpts is available from: http://climatecasechart.com/non-us-case/future-generation-v-ministry-environment-others/ [accessed 5 February 2020].
80 Kotzé, 2016, p. 56.
81 The role of the Charter is here particularly significant, since the EUCJ case law has insisted in its imperative effect, as, for example, in Cases C-569/16 and C-570/16, Stadt Wuppertal v Maria Elisabeth Bauer and Volker Willmeroth v Martina Broßonn, 6.10.2018. ECLI:EU:C:2018:871. See Cienfuegos Mateo, 2019, p. 9.
82 De Cabo Martín, 2014, p. 59.
83 Cienfuegos Mateo, 2019, p. 13.

References

Adelman, S., 2010. Climate Change, Human Rights and Corporate Accountability. In Humphreys, S. (Editor), *Human Rights and Climate Change*. Cambridge and New York: Cambridge University Press, 159–179.

Alder, J. and Wilkinson, D., 1999. *Environmental Law & Ethics*. London: Macmillan.

Alexy, R., 2003. Los derechos fundamentales en el Estado constitucional democrático. In Carbonell, M. (Editor), *Neoconstitucionalismo(s)*. Madrid: Trotta, 31–47.

Allen, M.R. and Lord, R., 2004. The Blame Game. *Nature*, 432, 551–552.

Anand, R., 2004. *International Environmental Justice*. Aldershot and Burlington: Ashgate 2004.

Anderson, D. and Murphy, C.C., 2012. The Charter of Fundamental Rights. In Biondi, A., Feckhart, P. and Ripley, S. (Editors), *EU Law after Lisbon*. Oxford: Oxford University Press, 155–179.

Anthropocene Working Group, 2019. *Results of binding vote by AWG—Released 21st May 2019*. Available from: http://quaternary.stratigraphy.org/working-groups/anthropocene/ [Accessed 21 August 2020].

Antypas, A. et al., 2009. Linking Environmental Protection, Health, and Human Rights in the European Union: An Argument in Favour of Environmental Justice Policy. *Environmental Law & Management*, 20, 8–21.

Atapattu, S., 2019. Environmental Justice, Climate Justice and Constitutionalism: Protecting Vulnerable States and Communities. In Jaria-Manzano, J. and Borràs, S. (Editors), *Research Handbook on Global Climate Constitutionalism*. Cheltenham and Northampton: Edward Elgar Publishing, 195–215.

Azar, C. and Holmberg, J., 1995. Defining the Generational Debt. *Ecological Economics*, 14(1), 7–19.

Baskin, J., 2019. Global Justice and the Anthropocene: Reproducing a Development Story. In Biermann, F. and Lövbrand, E. (Editors), *Anthropocene Encounters. New Directions in Green Political Thinking*. Cambridge, New York, Melbourne and New Delhi: Cambridge University Press, 150–168.

Bonneuil, C. and Fressoz, J.-B., 2017. *The Shock of the Anthropocene: The Earth, History and Us*. English edition by David Fernbach. London and Brooklyn: Verso.

Bosselmann, K., 2008. *The Principle of Sustainability. Transforming Law and Governance*. Farnham and Burlington: Ashgate.

Brunnée, J., 2009. Climate Change, Global Environmental Justice and International Environmental Law. In Ebbeson, J. and Okowa, P. (Editors), *Environmental Law and Justice in Context*. Cambridge: Cambridge University Press, 316–332.

Burke, A. and Fishel, S., 2019. Power, Politics, and Thing-Systems in the Anthropocene. In Biermann, F. and Lövbrand, E. (Editors), *Anthropocene Encounters. New Directions in Green Political Thinking*. Cambridge, New York, Melbourne and New Delhi: Cambridge University Press, 87–108.

Canosa Usera, R., 1996. Aspectos Constitucionales del Derecho Ambiental. *Revista de Estudios Políticos*, 94, 73–109.

Castree, N., 2019. The "Anthropocene" in Global Change Science: Expertise, the Earth and the Future of Humanity. In Biermann, F. and Lövbrand, E. (Editors), *Anthropocene Encounters. New Directions in Green Political Thinking*. Cambridge, New York, Melbourne and New Delhi: Cambridge University Press. 25–49.

Cienfuegos Mateo, M., 2019. El paper actual del Tribunal de Justícia de la Unió Europea en la governança de la Unió. *Revista Catalana de Dret Públic*, 59, 1–20.

Clarkson, S. and Wood, S., 2009. *A Perilous Imbalance. The Globalization of Canadian Law and Governance*. Vancouver and Toronto: University of British Columbia Press.

Cole, L.W. and Foster, S.L., 2001. *From the Ground Up. Environmental Racism and the Rise of Environmental Justice Movement*. New York: New York University Press.

Cortina, A., 2002. *Por una ética del consumo*. Madrid: Taurus.

Crutzen, P.J., 2002. Geology of Mankind. *Nature*, 415, 23.

Crutzen, P.J. and Stoermer, E.F., 2000. The "Anthropocene". *Global Change Newsletter*, 41, 17–18.

Cutter, Susan L., 1995. Race, Class and Environmental Justice. *Progress in Human Geography*, 19(1), 111–122.

Dal Bó, E., 2006. Regulatory Capture: A Review. *Oxford Review of Economic Policy*, 22(2), 203–225.

De Cabo Martín, C., 2014. *Pensamiento crítico, constitucionalismo crítico*. Madrid: Trotta.

Derman, B.D., 2019. Revisiting Limits to Legal Mobilization for Global Climate Justice: Complexity, Territoriality, and Responsibility. *Oñati Socio-legal Series*, 9(3), 333–360.

Dernbach, J. and Cheever, F., 2015. Sustainable Development and Its Discontents. *Transnational Environmental Law*, 4(2), 247–287.

Dobson, A., 1998. *Justice and the Environment. Conceptions of Environmental Sustainability and Dimensions of Social Justice*. Oxford: Oxford University Press.

Ebbeson, J., 2009. Introduction: Dimensions of Justice in Environmental Law. In Ebbeson, J. and Okowa, P. (Editors), *Environmental Law and Justice in Context*. Cambridge: Cambridge University Press, 1–36.

Escribano Collado, P., 1991. Ordenación del territorio y medio ambiente en la Constitución. In Martín-Retortillo, S. (Editor), *Estudios sobre la Constitución Española. Homenaje al profesor Eduardo García de Enterría (IV)*. Madrid: Civitas, 3705–3750.

Ewald, S., 2011. Adjudication of the Right to Education and the Right to Welfare. *Columbia Journal of Environmental Law*, 36(2), 413–459.

Falk, R., 2009. The Second Cicle of Ecological Urgency: An Environmental Justice Perspective. In Ebbeson, J. and Okowa, P. (Editors), *Environmental Law and Justice in Context*. Cambridge: Cambridge University Press, 39–54.

Figueroa, R.M., 2011. Indigenous Peoples and Cultural Losses. In Dryzek, J.S., Norgaard, R.B. and Schlosberg, D. (Editors), *The Oxford Handbook of Climate Change and Society*. Oxford: Oxford University Press, 231–247.

Fischer-Kowalski, M., 1998. Society's Metabolism: The Intellectual History of Materials Flow Analysis, Part I, 1860–1970. *Journal of Industrial Ecology*, 2, 61–78.

Fischer-Kowalski, M. and Hüttler, W., 1998. Society's Metabolism: The Intellectual History of Materials Flow Analysis, Part II, 1980–1998. *Journal of Industrial Ecology*, 2, 107–136.

Foster, S., 1999. Public Participation. In Gerrard, M.B. (Editor), *The Law of Environmental Justice. Theories and Procedures to Address Disproportionate Risks*. Chicago: American Bar Association, 185–229.

Gaja, G., 2012. Accession to the ECHR. In Biondi, A., Feckhart, P. and Ripley, S. (Editors), *EU Law after Lisbon*. Oxford: Oxford University Press, 180–194.

Gauna, E., 1995. Federal Environmental Citizen Provisions: Obstacles and Incentives on the Road to Environmental Justice. *Ecology Law Quarterly*, 22(1), 1–87.

Gellers, J.C. and Cheatham, T.J., 2019. Sustainable Development Goals and Environmental Justice: Realization through Disaggregation? *Wisconsin International Law Journal*, 36(2), 276–297.

Gerlach, J.W., 1989. *Privatrecht und Umweltschutz im System des Umweltrechts*. Berlin: Duncker & Humblot.

Girardi, G., 2014. Pueblos indígenas, ecologismo político y religión. *Papeles de relaciones ecosociales y cambio global*, 125, 125–137.

González, Carmen G., 2015. Environmental Justice, Human Rights, and the Global South. *Santa Clara Journal of International Law*, 13, 151–195.

Grear, A., 2010. *Redirecting Human Rights. Facing the Challenge of Corporate Legal Humanity*. Basingstoke and New York: Palgrave Macmillan.

Hiskes, R.P., 2009. *The Human Right to a Green Future. Environmental Rights and Intergenerational Justice*. New York: Cambridge University Press.

Hornborg, A., 2009. Zero-Sum World. Challenges in Conceptualizing Environmental Load Displacement and Ecologically Unequal Exchange in the World-System. *International Journal of Comparative Sociology*, 50(3–4), 237–262.

Jaria-Manzano, J., 2012. Environmental Justice, Social Change and Pluralism. *IUCN Academy of Environmental Law e-Journal*, 1, 18–29.

Jaria-Manzano, J., 2019. Law in the Anthropocene. In Jaria-Manzano, J. and Borràs, S. (Editors), *Research Handbook on Global Climate Constitutionalism*. Cheltenham and Northampton: Edward Elgar Publishing, 31–49.

Jaria-Manzano, J. and Borràs, S., 2019. Introduction the 'Research Handbook on Global Climate Constitutionalism. In Jaria-Manzano, J. and Borràs, S. (Editors), *Research Handbook on Global Climate Constitutionalism*. Cheltenham and Northampton: Edward Elgar Publishing, 1–16.

Jaria-Manzano, J. et al., 2016. Measuring Environmental Injustice: How Ecological Debt Defines a Radical Change in the International Legal System. *Journal of Political Ecology*, 23, 381–393.

Jositsch, D., 1997. Das Konzept der nachhaltigen Entwicklung (Sustainable Development) im Völkerrecht und seine innerstaatliche Umweltsetzung. *Umweltrecht in der Praxis/Le Droit de l'environnement dans la pratique*, 93–121.

Karpen, U., 1988. Zu einem Grundrecht auf Umweltschutz. In Thieme, W. (Editor), *Umweltschutz im Recht*. Berlin: Duncker & Humblot, 9–24.

Kjellén, B., 2009. Justice in Global Environmental Negotiations: The Case of Desertification. In Ebbeson, J. and Okowa, P. (Editors), *Environmental Law and Justice in Context*. Cambridge: Cambridge University Press, 333–347.

Kotzé, L.J., 2016. *Global Environmental Constitutionalism in the Anthropocene*. Oxford and Portland: Hart Publishing.

Lazarus, R.J., 1994. Pursuing "Environmental Justice": The Distributional Effects of Environmental Protection. *Land Use and Environment Law Review*, 263–333.

Llano, A., 1988. *La nueva sensibilidad*. Madrid: Espasa.

Malm, A. and Hornborg, A., 2014. The Geology of Mankind? A Critique of the Anthropocene Narrative. *The Anthropocene Review*, 1(1), 62–69.

Marquet Sardà, C., 2010. *Los derechos sociales en el Ordenamiento jurídico sueco. Estudio de una categoría normativa*. Barcelona: Atelier.

Martínez-Alier, J., 2002. The Ecological Debt. *Kurswechsel*, 4, 5–16.

Mary Robinson Foundation, n.d. *Principles of Climate Justice*. Available from: www.mrfcj.org/principles-of-climate-justice/ [Accessed 1 February 2020].

McWilliams, D.A., 1994. Environmental Justice and Industrial Redevelopment: Economics and Equality in Urban Revitalization. *Ecology Law Quarterly*, 21(3), 705–783.

Mesa Cuadros, G., 2011. Elementos para una teoría de la justicia ambiental. In Mesa Cuadros, G. (Editor), *Elementos para una teoría de la Justicia Ambiental y el Estado Ambiental de Derecho*. Bogotá: Universidad Nacional de Colombia, 25–62.

Mickelson, K., 2007. Critical Approaches. In Bodansky, D., Brunnée, J. and Hey, E. (Editors), *The Oxford Handbook of International Environmental Law*. Oxford: Oxford University Press, 262–290.

Mickelson, K., 2009. Competing Narratives of Justice in North-South Environmental Relations: The Case of Ozone Layer Depletion. In Ebbeson, J. and Okowa, P. (Editors), *Environmental Law and Justice in Context*. Cambridge: Cambridge University Press, 297–315.

Murswiek, D., 1995. *Umweltschutz als Staatszweck*. Bonn: Economica.

Nogueira López, A., 2011. Crisis económica y cambios estructurales en el régimen de ejercicio de actividades. In Blasco Esteve, A. (Editor), *El Derecho público de la crisis económica. Transparencia y sector público. Hacia un nuevo Derecho administrativo*. Madrid: INAP, 121–187.

Paredis, E. et al., 2004. *Elaboration of the Concept of Ecological Debt*. Ghent: Centre for Sustainable Development (CDO), Ghent University.

Schlosberg, D., 2004. Reconceiving Environmental Justice: Global Movements and Political Theories. *Environmental Politics*, 13(3), 517–540.

Serrano Moreno, J.L., 1992. *Ecología y Derecho: principios de Derecho Ambiental y Ecología jurídica*. Granada: Comares.

Steffen, W. et al., 2007. The Anthropocene: Are Humans Now Overwhelming the Great Forces of Nature? *Ambio*, 36(8), 614–621.

Taylor, P.J. and Flint, C., 2011. *Political Geography: World-Economy, Nation-State and Locality*, 6th ed. London: Routledge.

Thürer, D., 2001. Recht der internationalen Gemeinschaft und Wandel der Staatlichkeit. In Thürer, D., Aubert, J.F. and Müller, J.P. (Editors), *Verfassungsrecht der Schweiz/Droit constitutionnel suisse*. Zurich: Schulthess, 37–61.

Tridimas, T., 1999. *The General Principles of EC Law*. Oxford: Oxford University Press.

United States Environmental Protection Agency, n.d. *Environmental Justice*. Available from: https://www.epa.gov/environmentaljustice [Accessed 29 January 2020].

Untawale, M.G., 1990. Global Environmental Degradation and International Organizations. *International Political Science Review*, 11(3), 371–383.

Wallerstein, I., 1974. *The Modern World-System, vol. I: Capitalist Agriculture and the Origins of the European World-Economy in the Sixteenth Century*. New York and London: Academic Press.

Walters, M., 1991. Ecological Unity and Political Fragmentation: The Implications of the Brundtland Report for the Canadian Constitutional Order. *Alberta Law Review*, XXIX(2), 420–449.

Wapner, P., 2019. The Ethics of Political Research in the Anthropocene. In Biermann, F. and Lövbrand, E. (Editors), *Anthropocene Encounters. New Directions in Green Political Thinking*. Cambridge, New York, Melbourne and New Delhi: Cambridge University Press, 212–227.

Weisz, H., 2007. Combining Social Metabolism and Input-Output Analysis to Account for Ecologically Unequal Trade. In Hornborg, A., McNeill, J.R. and Martínez-Alier, J. (Editors), *Rethinking Environmental History: World-System History and Global Environmental Change*. Lanham: AltaMira Press, 289–306.

Westaway, J., 2012. Globalization, Transnational Corporations and Human Rights – A New Paradigm. *International Law Research*, 1(1), 63–72.

Wildhaber, L., 1972. Soziale Grundrechte. In Saladin, P. and Wildhaber, L. (Editors), *Der Staat als Ausgabe. Gedenkschrift für Max Imboden*. Basel and Stuttgart: Helbing & Lichtenhahn, 371–391.

World Commission on Environment and Development, 1987. *Report of the World Commission on Environment and Development: Our Common Future*. Available from: https://sustainabledevelopment.un.org/content/documents/5987our-common-future.pdf.

Index

Note: Page numbers in **bold** indicate a table, and page numbers followed by an 'n' indicate a note on the corresponding page.

Aarhus Convention 6, 93–102, 104n28, 105n55
access to justice 90–93; and Aarhus Convention 93–102
Agenda 2030 for Sustainable Development 8
Agreement on Port State Measures (PSMA) 73, 78, 81–82, 84
Aichi Targets 44
Antarctica 47n30
Antarctic Convention 38
Anthropocene 166–167, 171, 172
Anthropocene Working Group 175n9
Argentina 60–61
aviation 15, 18

Bern Convention on the Conservation of European Wildlife and Natural Habitats of the Council of Europe (Bern Convention) 134, **137**
biodiversity 3, 35–37; actions taken to protect 47n41; in Antarctica 47n30; and EU competence 37–39; and EU diplomacy 39–43; and Green Multilateralism 43–45; and indigenous people 46n23; and restoration of ecosystems 48n65
Biodiversity Financing Initiative (BIOFIN) 42
Biodiversity for Life Project 42
Biodiversity Summit 4
biofuels 61–62, 64
border taxes 51, 62–64
Brexit 4, 21–24, 26–27; *see also* United Kingdom (UK)
Bulgaria 124, 154

carbon dioxide 20, 29n56; *see also* greenhouse gases (GHG)
carbon footprint 60–63; *see also* greenhouse gases (GHG)
Cartagena Protocol on Biosafety to the Convention on Biological Diversity (Cartagena Protocol) 133, **137**
Catch Documentation Schemes (CDSs) 81
China 4, 12, 20–21, 25–26, 29n60; and carbon dioxide 29n56; and dumping 64; and Paris Agreement 19–21; and sustainable development 57–59
climate change 12–14; and Brexit 21–27; comprehensive approach to 16–19; and EU competences 14–16; and Paris Agreement 19–21; *see also* emissions trading; greenhouse gases (GHG)
Colombia 173, 178n79
commercial conservation and management measures (CMMs) 75
Common European Asylum System (CEAS) 120
Common Fisheries Policy (CFP) 75, 76–77, 84
competences 14–16, 37–39
Comprehensive Economic and Trade Agreement (CETA) 54
Conference of the Parties (CoP) 12
Conference of the Parties of the Convention on Biological Diversity (CBD CoP) 35–36
conservation 37, 39, 41, 42, 46n23, 56, 73–76, 80, 85n20, 86n53; *see also* biodiversity; fishing; timber
Convention on Biological Diversity

Convention on International Trade in
    Endangered Species of Wild Fauna and
    Flora (CITES Convention) 133, **137**
Convention on the Legal Trafficking of
    Endangered Species of Wild Flora and
    Fauna (CITES) 43
Copenhagen Conference 25
COVID-19 4, 10, 21, 26
criminal law 7; and Directive 2008/99/
    EC 136–138; and environmental
    protection 138–143; and multilateral
    environmental agreements (MEAs)
    132–136; *see also* environmental justice
Cyprus 124, 154

Denmark 29n66, 154, 158
diplomacy 37–43
Directive 2008/99/EC 7
dumping 56–60

Earth System Science 175n9
ecological debt 168, 172, 176n27
emissions trading 24; *see also* climate
    change; greenhouse gases (GHG)
environmental crimes: and Directive
    2008/99/EC 136–138; and
    environmental protection 138–143; and
    multilateral environmental agreements
    (MEAs) 133–136
Environmental Goods Agreement
    (EGA) 51
environmental justice 166–167, 176n24,
    176n27; and EU law 172–174; and
    fundamental values 170–172; vs
    sustainable development 167–170
EU Biodiversity Strategy 40, 43–45
EU-China Leaders' Statement on Climate
    Change and Clean Energy 21
EU Emissions Trading Scheme (EU ETS)
    12, 18–19, 26; *see also* greenhouse
    gases (GHG)
European Commission 9; and biodiversity
    37, 38, 40, 43; and climate change 15,
    16–18, 25; and environmental justice
    94–95, 99, 101–102; and environmental
    refugees 119–122; and fishing 76, 80;
    and sustainable development 50–51, 53,
    55, 59, 60, 157
European Convention on Human Rights
    (ECHR) 170
European Council 122–123
European External Action Service (EEAS)
    4, 14, 16, 40

European Fisheries Control Agency
    (EFCA) 77
European Free Alliance (EFA) 118
European General Framework for Judicial
    Training 93
European Green Deal (EGD) 51, 54; and
    access to justice 101; and biodiversity
    37; and climate change 5, 10, 17; and
    sustainable development 151, 153,
    155–159, 162n61, 162n63, 163n70
European Parliament 117–118
European Union Court of Justice (EUCJ)
    4, 15, 94–96
extinction 35–36

Finland 124, 125, 154, 158
fishing 5, 7, 73–84, 85n12, 85n20,
    138–140
Food and Agriculture Organization (FAO)
    73, 140
Forest Law Enforcement, Governance and
    Trade (FLEGT) 138

Generaciones Futuras 173, 178n78
General Agreement on Tariffs and Trade
    (GATT) 52
Generalised Scheme of Preferences (GSP)
    51, 52–53
Generalised System of Preferences 5
Geneva Convention Relating to the Status
    of Refugees 111
Global Biodiversity Framework of the
    Convention on Biological Diversity 4
Global Compact on Safe, Orderly and
    Regular Migration 112
gold 62
good governance 52–53
Greece 114
Green Climate Fund 20
Green Customs Initiative 143, 146n69
Green Deal *see* European Green Deal
    (EGD)
Green Diplomacy Network (GDN) 39,
    47n39
greenhouse gases (GHG) 3, 17–19, 26,
    61; and Brexit 22; and sustainable
    development 157; *see also* carbon
    dioxide; carbon footprint; climate
    change; emissions trading; EU
    Emissions Trading Scheme (EU ETS);
    European Green Deal (EGD)
Greenland 29n66
Green Multilateralism 37, 43–45

Habitats and Birds Directives 36
Havana Charter for an International Trade
    Organization 175n19
High Ambition Coalition (HAC) 13
human rights 50, 90, 102n1, 109–110, 116,
    125, 169–172; *see also* Kyoto protocol

illegal, unreported, and unregulated
    fishing (IUU) 5–6, 73–75, 87n63,
    138–140; and Agreement on Port States
    Measures (PSMA) 81–84; legal concept
    of 75–76; and non-cooperating third
    countries 79–81; regulations on 76–79
indigenous people 46n23
Indonesia 61
inequality 8, 166–167, 171, 172, 174n2;
    see also environmental justice
International Civil Aviation Organization
    (ICAO) 15
International Convention for the
    Prevention of Pollution from Ships
    (MARPOL Convention) 133, 134–136
International Labour Organization
    (ILO) 81
International Maritime Organization
    (IMO) 81
International Organization for Migration
    (IOM) 110
International Plan of Action to Prevent,
    Deter and Eliminate Illegal, Unreported
    and Unregulated Fishing (IPOA-IUU)
    74, 75–76
International Tribunal for the Law of the
    Sea (ITLOS) 76
Italy 114, 124, 125

Joint Declaration and Partnership on
    Climate Change of 2005 20
justice, access to 90–93; and Aarhus
    Convention 93–102
justice, environmental 166–167, 176n24,
    176n27; and EU law 172–174; and
    fundamental values 170–172; vs
    sustainable development 167–170

Korea, Republic of 48n59
Kyoto Protocol 12, 14, 30n87; and Trump,
    Donald 19

least-developed countries (LDCs) 52
Leyen, Ursula von der 17, 51, 54
Lisbon Treaty 16, 38, 44, 134; and Green
    Diplomacy Network (GDN) 40

Memoranda of Understanding 20
migration 6–7; policies on 113–117; *see
    also* refugees; refugees, environmental
Millenium Development Goals (MDGs) 150
Montreal Protocol on Substances that
    Deplete the Ozone Layer (Montreal
    Protocol) 134, **137**
multiannual financial framework
    (MFF) 155
multilateral environmental agreements
    (MEAs) 132–136, **137**

Nagoya Protocol 44
Nationally Determined Contributions
    (NDCs) 13, 23
natural resources 8; *see also* fishing; timber
Netherlands, the 173, 178n78
non-governmental organizations (NGOs)
    98–99
non-market economy (NME) 57–59

Obama, Barack 50

palm oil 61
Paris Agreement 4, 12, 13, 19–21,
    26, 30n76; and biodiversity 41; and
    environmental refugees 112; and Trump,
    Donald 19–20
poverty 74, 150, 169, 175n20; *see also*
    inequality
procurement 55–56

refugees 6–7; *see also* migration
refugees, environmental 109–110; and
    environmental change 110–113; and
    migration policy 113–117; protecting
    123–126; recognition of 117–123
regional fisheries bodies (RFBs) 74
regional fisheries management
    organizations 86n44
regional fisheries management
    organizations (RFMOs) 74, 75–76
Regional Protection Programmes
    (RPPs) 124
regional trade agreements (RTAs) 53–55
responsible fisheries 85n12
Romania 154
Rotterdam Convention on the Prior
    Informed Consent Procedure for Certain
    Hazardous Chemicals and Pesticides
    in International Trade (Rotterdam
    Convention) 133, **137**
Russia 59, 64

sanctions 133–136
Seventh Environment Action Programme
    (7EAP) 2; and biodiversity 38–40
social metabolism 166–169, 171–172,
    174n6
Spain 75, 114
steel 62
Stockholm Convention on Persistent
    Organic Pollutants (Stockholm
    Convention) 133, **137**
Stockholm Declaration on the Human
    Environment 90
sustainable development 50–52, 149–156;
    and carbon footprint 60–63; and
    dumping 56–60; and environmental
    justice 168–169; and EU leadership
    63–65; and European Green Deal
    156–159; and Generalised Scheme of
    Preferences (GSP) 52–53; and public
    procurement 55–56; and regional trade
    agreements (RTAs) 53–55
Sustainable Development Goals (SDGs)
    2, 7–8, 50, 149–159, 160n13, 160n23,
    161n34; and access to justice 91; and
    environmental justice 169, 173; and
    fishing 74
Sweden 124, 125, 154, 158
Syria 114

Temporary Protection Directive 115–116,
    118, 124, 126
timber 7, 138–139
trade: illegal 7; *see also* regional trade
    agreements (RTAs)
trade agreements 3, 5
Trade and Sustainable Development (TSD)
    54–55
Transatlantic Trade and Investment
    Partnership (TTIP) 50
transportation 10
Treaty of European Union (TEU) 7,
    and climate change 14–17; and
    environmental justice 93, 168–169, 170,
    171; environmental refugees 125; and
    sustainable development 50, 152, 155
Treaty on the Functioning of the European
    Union (TFEU) 1, 4, 7, 13, 14–18,
    27n22; and climate change 13, 14,
    15–18; and environmental crime

140–141; and environmental justice
    97–99, 100, 102; and environmental
    refugees 115; and fishing 83; and
    sustainable development 152
triple bottom line 156
Trump, Donald 19, 51

UN Conference on Environment and
    Development (UNCED) 150
Unique Vessel Identifiers (UVIs) 81
United Kingdom (UK) 21–24, 26–27
United Nations Agenda 2030 2
United Nations Conference on the Human
    Environment 149–150
United Nations Convention against
    Transnational Organized Crime
    (UNTOC) 141
United Nations Convention on the Law of
    the Sea (UNCLOS) 75, 76, 135
United Nations Convention to Combat
    Desertification (UNCCD) 110
United Nations Development
    Programme 42
United Nations Economic Commission for
    Europe (UNECE) 90
United Nations Environment Programme
    (UNEP) 36, 119
United Nations Framework Convention on
    Climate Change 30n76
United Nations Framework Convention on
    Climate Change (UNFCCC) 13–14; and
    Trump, Donald 19
United Nations High Commissioner for
    Refugees (UNHCR) 7, 110–111
United Nations (UN) 50
United States (US) 4, 12; and
    environmental justice 166; and Paris
    Agreement 19–21; and sustainable
    development 50
Urgenda 173, 178n78

Vienna Convention on the Law of Treaties
    (VCLT) 60

World Trade Organization (WTO) 41, 50;
    and dumping 56–59; and sustainable
    development 55–56

Xi Jinping 12